THERMODYNAMICS OF THE ATMOSPHERE: A COURSE IN THEORETICAL METEOROLOGY

This textbook is written for graduate students and researchers in meteorology and related sciences. To encourage the reader to follow in detail the mathematical developments, the derivations are complete leaving out only the most elementary steps. While most meteorological textbooks only present equilibrium thermodynamics, this book also introduces the linear theory of nonequilibrium and provides the necessary background for more advanced studies. It is designed to complement the authors' previous text, *Dynamics of the Atmosphere* (Cambridge, 2003).

The book starts with an introduction to the basic concepts, including the general form of the budget equation. The authors show how to derive the prognostic equations for temperature that follow from the development of the first law of thermodynamics. They then go on to discuss the second law of thermodynamics. The entropy production equation is introduced on the basis of Gibbs' fundamental equation and the calculation of entropy changes is demonstrated. A brief treatment of the thermodynamics of blackbody radiation is followed by a detailed discussion of thermodynamic potentials. Of great importance are the constitutive equations of irreversible fluxes, which are described in connection with the Onsager–Casimir reciprocity relations. Later chapters discuss the state functions of ideal gases, of the condensed pure phase, and the state functions, as well as the thermodynamics of cloud air. The heat equations for special adiabatic systems are described and the authors introduce the use of graphical procedures to evaluate atmospheric variables. The final chapter presents essential topics of atmospheric statics making use of such concepts as homotropy, barotropy, heterotropy, and piezotropy. This chapter contains a thorough discussion of stability and of atmospheric energetics of hydrostatic equilibrium.

Each chapter ends with a set of exercises that are designed to help the reader develop a deeper understanding of the subject. Answers to all the exercises are given at the end of the book.

WILFORD ZDUNKOWSKI received B.S. and M.S. degrees from the University of Utah and was awarded a Ph.D. in meteorology from the University of Munich in 1962. He then returned to the Department of Meteorology at the University of Utah, where he was later made Professor of Meteorology. In 1977, he took up a professorship, at the Universität Mainz, where for twenty years, he taught courses related to the topics presented in this book. Professor Zdunkowski has been the recipient of numerous awards from various research agencies in the USA and in

Germany, and has traveled extensively to report his findings to colleagues around the world.

ANDREAS BOTT received a Diploma in Meteorology from the University of Mainz, in 1982 and subsequently worked as a research associate under Professor Paul Crutzen at the Max-Planck-Institut für Chemie in Mainz, where he was awarded a Ph.D. in Meteorology in 1986. He held a variety of positions at the Institute for Atmospheric Physics in the Universität Mainz between 1986 and 1999, and during this time he also spent periods as a guest scientist at institutions in the USA, Norway, and Japan. Since 2000, Dr Bott has been a University Professor for Theoretical Meteorology at the Rheinische Friedrich-Wilhelms-Universität in Bonn. Professor Bott teaches courses in theoretical meteorology, atmospheric thermodynamics, atmospheric dynamics, cloud microphysics, atmospheric chemistry and numerical modeling.

THERMODYNAMICS OF THE ATMOSPHERE
A COURSE IN THEORETICAL METEOROLOGY

WILFORD ZDUNKOWSKI
and
ANDREAS BOTT

CAMBRIDGE
UNIVERSITY PRESS

CAMBRIDGE UNIVERSITY PRESS
Cambridge, New York, Melbourne, Madrid, Cape Town, Singapore, São Paulo

Cambridge University Press
The Edinburgh Building, Cambridge CB2 8RU, UK

Published in the United States of America by Cambridge University Press, New York

www.cambridge.org
Information on this title: www.cambridge.org/9780521809535

First published 2004

A catalogue record for this publication is available from the British Library

Library of Congress Cataloguing in Publication data

Zdunkowski, Wilford, 1929–
Thermodynamics of the atmosphere/by Wilford Zdunkowski and Andreas Bott.
p. cm
Includes bibliographical references and index.
ISBN 0 521 80953 3 – ISBN 0 521 00685 6 (paperback)
1. Atmospheric thermodynamics. I. Bott, Andreas, 1956– II. Title.

QC880.4.T5Z38 2004
551.5'22–dc21   2003055188

ISBN 978-0-521-80953-5 hardback
ISBN 978-0-521-00685-9 paperback

Transferred to digital printing 2007

This book is dedicated to the memory of
K. H. Hinkelmann (1915–1986)
and G. Korb (1928–1991)
who excelled as theoretical meteorologists and
teachers of meteorology.

Contents

Preface

Thermodynamics of the Atmosphere is the second volume in the series *A Course in Theoretical Meteorology*. In the first volume, entitled *Dynamics of the Atmosphere*, we have covered many of the essential topics of atmospheric motion, but we did not provide any theory from thermodynamics. Whenever information from thermodynamics was required for the development of various topics of dynamic meteorology, we have carefully stated the basic facts without giving any theoretical background. The reader of this book will clearly recognize that *Dynamics of the Atmosphere* and *Thermodynamics of the Atmosphere* are so closely connected that one textbook without the other would give a very incomplete picture of atmospheric weather systems.

We have tried to make this book as self-contained as possible. Whenever some basic facts from atmospheric dynamics are required, we have simply stated these omitting any theoretical considerations. Occasionally, however, it might be desirable, to provide some additional background. In this case we make reference to our book *Dynamics of the Atmosphere*, which will be abbreviated by *DA*. Of course, other textbooks might be just as satisfactory, but it would be more difficult for the reader to extract the necessary information in the desired form.

Thermodynamics of the Atmosphere is written for advanced undergraduate and recently graduated students of meteorology and related disciplines. The book contains a sufficient amount of both theory and applications to provide a solid foundation for more advanced studies. This text is largely based on an excellent unpublished manuscript by K. H. Hinkelmann that we have used as the basis for our own lectures. However, we found it necessary to replace a number of topics by other material to make the book more versatile. The concluding chapter of this book gives a condensed but necessary survey of statics of the atmosphere making use of G. Korb's brief but concise synopsis of this subject.

We will now briefly summarize the major topics developed in this book. The first chapter presents various important definitions and concepts that are indispensible

for a concise description of atmospheric thermodynamics. In the second chapter
we discuss the general form of the budget equation since modern as well as classical topics of thermodynamics can be treated more adequately with the help of
this equation. The reader may be aware of the fact that many important equations
of meteorology and of other branches of physics can be written in budget form.
Chapter 3 presents the all-important first law of thermodynamics written in the
general budget form. This equation is transformed to give a concise and fairly complete form of the prognostic equation of atmospheric temperature. Various fluxes
appear in this equation that may be parameterized with the help of the second law of
thermodynamics, which is discussed in the fourth chapter. The concept of entropy
is introduced together with a fairly thorough discussion of the entropy production
and Gibbs' fundamental equation. Thermal radiation is treated briefly in Chapter 5
and the relationship between entropy and blackbody radiation is derived. Chapter 6
introduces the thermodynamic potentials. Various important identities, indispensable in our work, are derived for later use. Morever, the concept of thermodynamic
stability is introduced. Chapter 7 discusses some important parts of the theory of
irreversible processes. There it is shown how to construct the analytic form of the
superimposed atmospheric fluxes. The theory involves the important principle of
Curie and the Onsager–Casimir reciprocity relations. This chapter provides the basis for the parameterization of the microturbulent fluxes as discussed in Chapter 11
of *DA*. Chapters 8 and 9 provide detailed descriptions of the state functions of ideal
gases and the condensed pure phase. The combination of these functions results in
the state functions of cloud air as discussed in the tenth chapter. Some typical meteorological topics are treated in the following three chapters. Chapter 11 describes
the theory of thermodynamic filtering in order to introduce special adiabatic systems. Chapter 12 discusses special homogeneous systems and adiabats. Chapter 13
describes briefly some thermodynamic diagrams that can be used to determine
graphically various temperatures and the stability of atmospheric systems. At the
conclusion of this book we present a synopsis of some important topics of atmospheric statics. Concepts such as homotropy, heterotropy and piezotropy, together
with interesting special cases are discussed. A fairly thorough description of the
hydrostatic equilibrium is presented together with the atmospheric energetics of
this particular state.

   At the end of each chapter several problems are given with answers provided at
the end of the book. Some of these problems are little more than trivial and serve
the purpose to familiarize the student with new ideas and equations. Some other
problems are designed to test the student's ingenuity. Occasionally hints are added
to remove some stumbling blocks. To a large degree these problems were presented
in exercise classes to the meteorology students of the University of Mainz. We
were fortunate to be assisted by some very capable instructors to conduct these

exercise classes. We wish to express our sincere gratitude to Drs J. Eichhorn, G. Korb, R. Schrodin, J. Siebert and T. Trautmann. It would be impossible to name all contributors. Our special gratitude goes to Dr. W.-G. Panhans for his splendid cooperation with the authors in organizing and conducting many of these classes. Whenever asked, he also taught various courses on topics of theoretical meteorology in order to lighten our burden.

Most of all, the authors are grateful to Professor K. H. Hinkelmann and Dr G. Korb whose lecture notes provided a solid basis upon which to write this book. If they were still alive, we are sure, they would be pleased that this book has been written. In fact, they encouraged us to undertake the arduous task of writing this text.

Finally, we wish to thank our families for their constant support and encouragement.

It seems to be one of the unfortunate facts of life that no book as technical as this can be published free of error. However, we take some comfort in the thought that any errors appearing in the text are due to the co-author. To eventually remove these errors, we invite the readers of this book to point out to us any misprint and other mistakes they discover.

1

Basic thermodynamic concepts

The task ahead is the formulation of prognostic and diagnostic equations to describe the future development of the atmospheric thermodynamic state. The variables describing this state are also called the *state variables* of the thermodynamic system. These quantities may be the pressure, temperature and density of the air as well as the concentrations of the water substance in its different phases, that is water vapor, liquid water, and ice. As will be later seen, other choices of state variables are also possible.

The laws of atmospheric motion, as far as they are required in this book, are considered to be known. If necessary, additional details will be given at the appropriate places. A suitable reference book for our purposes is *Dynamics of the Atmosphere* by the present authors, which from now on will be abbreviated by *DA*. Of course, any other suitable textbook on atmospheric dynamics might be just as satisfactory.

1.1 Description of the atmospheric thermodynamic state

The physical object to be investigated is called the *thermodynamic system*, anything interacting with the system is defined as the *surroundings*. The thermodynamic state defines all properties of the system that have to be determined from measurements or from calculations. Furthermore, the system is assumed to satisfy the following conditions:

(i) The system may have an arbitrary size and shape. However, it must be large enough so that direct molecular interactions do not have to be treated explicitly. All processes are described in terms of the so-called *bulk variables*. Microphysical properties such as the curvature of small water droplets are not accounted for.
(ii) In case that the system contains liquid water and ice, the two substances are assumed to be completely separated so that mixed particles (graupel, etc.) are not considered.

The demarcation of the system from its surroundings may be problematical in the atmosphere, and the isolation from its surroundings can be approximately realized

1

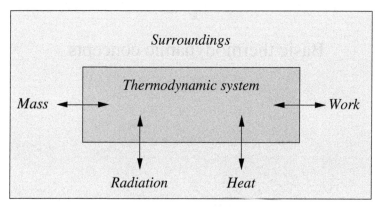

Fig. 1.1 Schematic representation of a thermodynamic system and its interaction with the surroundings.

Table 1.1. *Interaction of a thermodynamic system with its surroundings*

Type of interaction	Description of the system
No exchange of mass and energy	Completely closed or isolated
No exchange of mass, immaterial exchange (radiation, heat conduction) permitted	Closed with respect to mass
Exchange of mass and energy (heat in any form and work)	Open system

only when the interior interactions of the system dominate any interaction with the surroundings. If the thermodynamic system can be isolated from its surroundings for a sufficiently long period of time, then the system is said to approach an equilibrium state. Possible interactions of the system with its surroundings may be schematized as shown in Figure 1.1. As will be seen, there are various types of interactions possible which are listed in Table 1.1. Special open and closed systems will be described later when the appropriate terminology has been introduced.

If one observes a change in the system between times $t_2 > t_1$, some process has taken place. If the original state at t_1 cannot be restored without a lasting change within the surroundings, the process is called *irreversible*. In principle, all realizable processes are irreversible. An idealization is the so-called *reversible process* that is described by a series of equilibrium states, i.e. the system is in equilibrium at any time and, therefore, can be restored to its original state leaving no changes in the surroundings. This type of process may be approximated by a series of so-called *quasi-static processes*. In this case the state variables are independent of time.

Before further describing the quasi-static process, we discuss the concept of *thermodynamic equilibrium*. Such an equilibrium exists when the following stringent requirements are satisfied:

(i) *Mechanical equilibrium*—The force exerted by the system is uniform throughout the system and is completely balanced by *external forces*.
(ii) *Thermal equilibrium*—The temperature is uniform throughout the system and is the same as that of the surroundings.
(iii) *Chemical equilibrium*—The internal structure and the chemical composition of the system remain unchanged.

If the system happens to be in thermodynamic equilibrium and the conditions of the surroundings remain unchanged, no motion will occur and no *work* will be performed. If the sum of the external forces is changed for some reason, then a finite *unbalanced force* will be acting on the system and the conditions of mechanical equilibrium cannot be sustained. This may result in the production of turbulence and waves. These may disturb the uniform temperature distribution by introducing a finite temperature difference between the system and its surroundings. A nonuniform temperature may disturb the chemical equilibrium. The result is that the system will pass through nonequilibrium states. If it is desired during a process to describe every state of the complete system by means of thermodynamic variables, the process must not be caused by a finite unbalanced force. This leads us to assume an ideal situation in which external forces acting on the system are varied so slightly that the unbalanced force is infinitesimally small. In this case an equation of state is valid for all intermediary states. Any process caused by an infinitesimally small unbalanced force is called quasi-static meaning that the system is always infinitesimally close to thermodynamic equilibrium. The reversible process can be described by a succession of such states. The concept of thermodynamic equilibrium will also be utilized in the chapter on radiative transfer.

1.2 Basic definitions and terminology

In the following the necessary terminology will be introduced for an efficient mathematical description of the various atmospheric processes.

1.2.1 Composition of the air

The atmosphere is assumed to consist of two substances, the *dry air* and water in its three phases. Other substances such as aerosol particles are not accounted for. The dry air is treated as a constant mixture of different gases with nitrogen and oxygen being the main constituents. Details about the composition of dry air may be extracted from Table 8.1 of Chapter 8.

In order to efficiently manipulate the atmospheric variables, the following notation will be introduced for the mass components contained in an arbitrary atmospheric volume element:

M^0:   mass of dry air M^1:   mass of water vapor
M^2:   mass of liquid water M^3:   mass of ice

The sum of M^0 and M^1 is the mass of the *moist air*. As already mentioned, the masses M^2 and M^3 are bulk quantities, size distributions of cloud droplets or ice particles are not accounted for. Summing over all k results in the total air mass M.

For many purposes it is useful to operate with *mass concentrations* defined by the relations

$$m^k = \lim_{M \to 0} \frac{M^k}{M} \quad \text{with} \quad M = \sum_{k=0}^{3} M^k, \qquad \sum_{k=0}^{3} m^k = 1 \qquad (1.1)$$

The volume of each phase will be defined by the functional relations

$$
\begin{aligned}
V^0 &= V^0(p, T, M^0) & &\text{volume of dry air} \\
V^{0,1} &= V^{0,1}(p, T, M^0, M^1) & &\text{volume of moist air} \\
V^2 &= V^2(p, T, M^2) & &\text{volume of bulk liquid water} & \quad (1.2) \\
V^3 &= V^3(p, T, M^3) & &\text{volume of bulk ice} \\
V &= V^{0,1} + V^2 + V^3 & &\text{total volume}
\end{aligned}
$$

where p and T are the pressure and temperature of the air. If the system consists only of dry air and water vapor, then m^1 is equivalent to the *specific humidity* usually denoted by q.

1.2.2 Intensive and extensive variables

Imagine a system in equilibrium that will be divided into two equal parts, each containing the same mass. The properties of each half of the system that will remain unchanged are called intensive. Examples are pressure and temperature. The properties that are halved are called extensive. Obvious examples are mass and volume.

Before further clarifying the difference between intensive and extensive variables, we subdivide the total system consisting of dry air, water vapor, water and ice into numerous subsystems. If, regardless of the particular subdivision of the total system, the state variables m^k, p and T of each subsystem have the same values, then the total system is called *homogeneous*. Otherwise it is called *inhomogeneous*. Obviously the spatial gradients of all variables of state vanish in a

Water: V^2	Moist air: $V^{0,1}$	Ice: V^3
$\nabla m^2 = 0$	$\nabla m^0 = 0,$ $\nabla m^1 = 0$	$\nabla m^3 = 0$
	$\nabla p = 0,$ $\nabla T = 0$	

Fig. 1.2 Schematics of a homogeneous system.

homogeneous system. For visual purposes a homogeneous system may be represented as shown in Figure 1.2.

By definition, each individual phase is uniform in a homogeneous system. If the mass of each phase volume is multiplied by a factor $K > 0$, then the phase volume will be multiplied by the same factor providing that temperature and pressure gradients vanish as indicated in the figure. For example, if the liquid water mass is multiplied by the factor K then

$$V^2(p, T, K M^2) = K V^2(p, T, M^2) \qquad (1.3)$$

so that for the entire system

$$V(p, T, K M^0, K M^1, K M^2, K M^3) = K V(p, T, M^0, M^1, M^2, M^3) \quad (1.4)$$

Each thermodynamic function $\Psi(p, T, M^k)$ showing the same behavior as V is called an *extensive function*. The formal definition is given by

$$\boxed{\Psi(p, T, K M^k) = K^l \Psi(p, T, M^k) = K \Psi(p, T, M^k), \qquad l = 1, \ K > 0}$$
$$(1.5)$$

Referring to textbooks on mathematical analysis, Ψ will be recognized as a homogeneous function of first degree ($l = 1$) of the partial mass M^k. In contrast to this, an *intensive function* is defined by the relation

$$\boxed{\psi(p, T, K M^k) = K^l \psi(p, T, M^k) = \psi(p, T, M^k), \qquad l = 0, \ K > 0} \quad (1.6)$$

which is a homogeneous function of M^k of degree zero ($l = 0$). All extensive functions will be designated by a capital letter while intensive functions will not be capitalized. The only exception is the temperature T to avoid confusion with time t. Often it is useful to express the intensive quantity as a function of the concentration m^k by setting $K = 1/M$ in (1.6). In this case the intensive variable will be written as $\psi(p, T, m^k)$.

1.2.3 Specific extensive quantities

For a number of purposes it is useful to divide the extensive quantity Ψ of a homogeneous system by its total mass M or by the volume V or even by the number of moles N comprising the mass. The resulting quantities will be intensive. They are defined as

$$\psi = \frac{\Psi}{M} \qquad \text{specific value of } \Psi$$

$$\widehat{\psi} = \frac{\Psi}{V} \qquad \text{density of } \Psi \qquad (1.7)$$

$$\check{\psi} = \frac{\Psi}{N} \qquad \text{molar value of } \Psi$$

where N is the number of moles contained in M. In this text we will make little use of the molar value, but this definition is quite extensively used in textbooks on chemical thermodynamics. Some examples are the following specific extensive quantities

$$\Psi = M \implies \widehat{\psi} = \frac{M}{V} = \rho$$

$$\Psi = V \implies \psi = \frac{V}{M} = v = \frac{1}{\rho} \qquad (1.8)$$

$$\Psi = M^k \implies \psi = \frac{M^k}{M} = m^k, \qquad \widehat{\psi} = \frac{M^k}{V} = \rho^k$$

where ρ is the density and v is the specific volume of the air. The quantity ρ^k is known as the *partial density*, not to be confused with the *bulk density* $\rho_k = M^k/V^k$ where V^k is the volume occupied by M^k. For water the bulk density is $\rho_2 = 10^3 \text{ kg m}^{-3}$ while for a stratified cloud the partial density may be $\rho^2 = 10^{-4} \text{ kg m}^{-3}$. Furthermore, from (1.8) it is seen that $\widehat{\psi}$ can always be replaced by $\rho\psi$.

Finally, the following expressions are easy to verify:

$$\sum_{k=0}^{3} m^k = 1 \implies \sum_{k=0}^{3} dm^k = 0, \qquad \sum_{k=0}^{3} \rho^k = \rho \qquad (1.9)$$

It should be noted that the coordinates or variables of state of the homogeneous system p, T and M^k may now be replaced by the intensive coordinates or variables p, T and m^k. Pressure and temperature are usually referred to as *physical variables* while M^k and m^k are known as *chemical variables*. Functions of the coordinates of state $\Psi(p, T, M^k)$ or $\psi(p, T, m^k)$ are called *state functions*. Heat added to the system and work done on or by the system are not state functions.

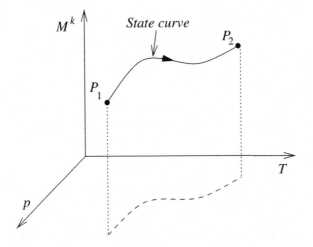

Fig. 1.3 Coordinates of state, state curve.

1.3 Description of the state of homogeneous systems

1.3.1 State coordinates

The thermodynamic state of homogeneous systems is considered to be known if the variables p, T and M^k describing the system of volume V can be specified. Additional information such as the velocity of the system and the shape of the volume is not needed to describe the thermodynamic state. Examples of state functions are the state coordinates themselves, the volume, the mass, the thermodynamic potentials (to be defined later) and various other quantities. For an ideal gas the state function v is specified by the ideal gas law

$$pv = RT \quad \text{or} \quad pV = NR^*T \tag{1.10}$$

where T is the absolute temperature, and R is the *individual gas constant*, i.e. the *universal gas constant* R^* divided by the molecular weight of the gas. Equation (1.10) is well known but will be derived later.

In a Cartesian coordinate system with the six axes p, T, M^k, $k = 0, 1, 2, 3$, the thermodynamic state of a homogeneous system is uniquely determined by a point, say P_1, see Figure 1.3. If the state of the system changes with time due to interactions with the surroundings of the system, or perhaps due to internal changes resulting from phase changes of the water substance, this point begins to move along the so-called *state curve* to a new state P_2. For the description of the thermodynamic state of the system at P_2 no information is needed about the state curve between P_1 and P_2.

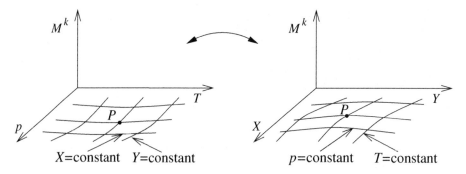

Fig. 1.4 Transformation of coordinates.

For the solution of specific problems it may be necessary to change coordinates. In place of (p, T) it may be of some advantage to use some other state functions, say X and Y. Let us introduce the two state functions

$$X = X(p, T, M^k), \qquad Y = Y(p, T, M^k) \qquad (1.11)$$

To demonstrate this change of variables, consider an arbitrary (p, T)-plane by fixing the coordinates M^k. A variation in M^k means the transition to another (p, T)-plane. In the (p, T)-plane the quantities $X = $ constant and $Y = $ constant generally result in families of curves and not of straight lines, see Figure 1.4. In this plane every state, represented by the point P, is associated with a unique pair (X_P, Y_P).

If the Jacobian, also known as functional determinant of the transformation between (X, Y) and (p, T), differs from zero,

$$J\left(\frac{X, Y}{p, T}\right) = \frac{\partial(X, Y)}{\partial(p, T)} = \begin{pmatrix} \dfrac{\partial X}{\partial p} & \dfrac{\partial X}{\partial T} \\ \dfrac{\partial Y}{\partial p} & \dfrac{\partial Y}{\partial T} \end{pmatrix} \neq 0 \qquad (1.12)$$

then to every pair (X, Y) a unique association with the pair (p, T) is guaranteed. Instead of (p, T) one may use, for example, the physical coordinate pairs (V, T), (p, S), (V, S) where S is some other variable (entropy) whose properties will be defined later. Whenever possible, we should stick to the original coordinates (p, T) since these variables are directly observed in the atmosphere. Furthermore, the extensive quantities are homogeneous functions of the partial masses M^k only in terms of the original variables (p, T).

1.3.2 Euler's theorem

Before continuing our discussion, we wish to introduce a special notation to avoid the writing of summation signs. Whenever a letter $m, n, p, q, \ldots$ is repeated in a term, we sum over this index from $0 - 3$. This technique is known as *Einstein's summation convention*. In contrast to the summation indices, the letters i, j, k, l are considered 'free' indices which are used to enumerate equations. Note that summation is not implied even if a free index occurs twice in a term. To give a specific example, we write $\sum_{k=0}^{3} \psi_k m^k = \psi_n m^n$. As a matter of convention, we are going to attach a subscript to one symbol and a superscript to the other symbol.

The definition of intensive and extensive state functions as given by equations (1.5) and (1.6) can be combined to the single statement for homogeneous functions of degree zero and one

$$K^l \Psi(p, T, M^k) = \Psi(p, T, K M^k) \qquad l = 0, 1 \qquad (1.13)$$

Much of the usefulness of these definitions results from the existence of Euler's well-known theorem from analysis which can be easily obtained from (1.13). To derive this theorem it is only necessary to differentiate this expression with respect to K for fixed p and T. The resulting expression must be valid for any value of K. Setting $K = 1$ yields Euler's theorem for homogeneous functions of degree l

$$l K^{l-1} \Psi(p, T, M^k) = \frac{\partial \Psi}{\partial (K M^n)} \frac{d(K M^n)}{dK}$$
$$K = 1 \implies l \Psi(p, T, M^k) = \frac{\partial \Psi}{\partial M^n} M^n \qquad (1.14)$$

For intensive and extensive state functions $l = 0$ and $l = 1$, respectively. Higher l-values are of little interest in thermodynamics. Equation (1.14) has been written with the help of the Einstein summation convention.

The partial derivatives of Ψ with respect to M^k are given the special name *partial specific property*

$$\psi_k = \left(\frac{\partial \Psi}{\partial M^k} \right)_{p,T} \qquad (1.15)$$

The interpretation of this quantity will be taken care of later. For the present ψ_k is a useful shorthand notation permitting an efficient representation of the homogeneous functions

$$\Psi = \psi_n M^n \quad \text{or} \quad \psi = \psi_n m^n \quad \text{or} \quad \rho \psi = \psi_n \rho^n \qquad (1.16)$$

Inspection of this equation shows that ψ_k depends on the same variables as ψ

$$\psi_k = \psi_k(p, T, m^l) \qquad (1.17)$$

Hence, ψ_k is homogeneous of degree zero. This also follows by differentiating $\Psi = \psi_n M^n$ with respect to M^k.

What is really of interest from the meteorological point of view is not the function itself but its variation. Let us consider a system with constant total mass M. By forming the total derivatives of $\Psi(p, T, M^k)$ and $\psi(p, T, m^k)$ we obtain

$$
\begin{aligned}
d\Psi &= d\left(\psi_n M^n\right) = \psi_n dM^n + M^n d\psi_n \\
&= \left(\frac{\partial \Psi}{\partial p}\right)_{T, M^k} dp + \left(\frac{\partial \Psi}{\partial T}\right)_{p, M^k} dT + \psi_n dM^n \\
d\psi &= d\left(\psi_n m^n\right) = \psi_n dm^n + m^n d\psi_n \\
&= \left(\frac{\partial \psi}{\partial p}\right)_{T, m^k} dp + \left(\frac{\partial \psi}{\partial T}\right)_{p, m^k} dT + \frac{\partial \psi}{\partial m^n} dm^n
\end{aligned}
\tag{1.18}
$$

From these equations we easily see that

$$
\begin{aligned}
M^n d\psi_n &= \left(\frac{\partial \Psi}{\partial p}\right)_{T, M^k} dp + \left(\frac{\partial \Psi}{\partial T}\right)_{p, M^k} dT \\
m^n d\psi_n &= \left(\frac{\partial \psi}{\partial p}\right)_{T, m^k} dp + \left(\frac{\partial \psi}{\partial T}\right)_{p, m^k} dT
\end{aligned}
\tag{1.19}
$$

Because of (1.9) one of the dm^k appearing in (1.18) can be eliminated, say dm^1. This gives

$$
\boxed{d\psi = \left(\frac{\partial \psi}{\partial p}\right)_{T, m^k} dp + \left(\frac{\partial \psi}{\partial T}\right)_{p, m^k} dT + \sum_{k=0}^{3} (\psi_k - \psi_1) dm^k}
\tag{1.20}
$$

from which follows directly

$$
\left(\frac{\partial \psi}{\partial m^k}\right)_{p, T} = \psi_k - \psi_1, \qquad k \neq 1
\tag{1.21}
$$

Since any one of the four terms dm^k can be eliminated, we write in general

$$
\boxed{\left(\frac{\partial \psi}{\partial m^k}\right)_{p, T} = \psi_k - \psi_i, \qquad k \neq i}
\tag{1.22}
$$

In order to obtain time changes we divide all total differentials in (1.19) or (1.20) by dt.

So far the discussion has been quite formal in terms of the unspecified function Ψ or ψ. Later, these functions will be given the proper identification as needed in meteorological thermodynamics. The advantage of this treatment is that the formulas derived in this section are valid for a collection of functions thus eliminating much repetition in later chapters.

Before moving on to the next chapter, a few remarks about the specific partial property $\psi_k = \psi_k(p, T, m^l)$ will be appropriate. Consider momentarily a homogeneous system of a single substance in one phase such as liquid cloud water or ice. Such a state will be signified as a *pure phase*. To distinguish the pure phase from a *mixed phase* such as moist air, the additional superscript ° will be attached to it. Since in a pure phase the concentration of the substance is 1, ψ_k cannot depend on m^l so that only the functional dependency on p and T is left yielding

$$\psi_k = \overset{\circ}{\psi}_k(p, T), \qquad k = 2, 3 \tag{1.23}$$

The situation differs in case of moist air ($m^2 = m^3 = 0$) which is a mixed phase so that the specific partial property depends in addition to p and T also on m^0 and m^1. However, m^0 may be eliminated in favor of $m^1 = 1 - m^0$ which for moist air is identical to the specific humidity q. Hence, for moist air we obtain

$$\psi_k = \psi_k(p, T, q), \qquad k = 0, 1 \tag{1.24}$$

In the special case of dry air ($q = 0$) or water vapor only ($q = 1$) we may write $\psi_0 = \overset{\circ}{\psi}_0$ and $\psi_1 = \overset{\circ}{\psi}_1$, respectively. Dry air is treated as a pure phase since its basic concentrations are constant for long periods of time. For mixed phases three different situations may occur

$$\psi_k \gtreqless \overset{\circ}{\psi}_k \tag{1.25}$$

To demonstrate that ψ_k and $\overset{\circ}{\psi}_k$ may differ indeed, we will give a practical example.

Consider a sufficiently large volume of a mixed phase consisting of water and alcohol, each part having the same number of moles. We add to this mixture 36×10^{-3} kg or 36×10^{-6} m^3 pure water, i.e.

$$\Delta M = 36 \times 10^{-3} \text{ kg}, \qquad \Delta \overset{\circ}{V} = 36 \times 10^{-6} \text{ m}^3, \qquad \overset{\circ}{v} = \frac{\Delta \overset{\circ}{V}}{\Delta M} = 10^{-3} \text{ m}^3 \text{ kg}^{-1} \tag{1.26}$$

The observed change in volume of the mixture is not 36×10^{-6} m^3, as one might expect, but only $\Delta V = 33 \times 10^{-6}$ m^3. This gives $v = \Delta V/\Delta M = 33/36 \times 10^{-3}$ m^3 kg$^{-1} < \overset{\circ}{v}$. Thus, it is concluded that each molecule of the alcohol and of the pure water requires more space in the pure phase than in the mixture.

In this chapter we have introduced some fundamental thermodynamic concepts such as the thermodynamic equilibrium, extensive and intensive quantities. We have briefly discussed Euler's theorem for homogeneous functions. This theorem led to the definition of the specific partial property ψ_k. Finally, we have derived the general differential expressions for the extensive and intensive variables Ψ and ψ.

1.4 Problems

1.1: Determine the degree l of the homogeneous function $V(r) = 4/3\pi r^3$, (a) with reference to the radius of the sphere, (b) with reference to the surface of the sphere. Solve the problem by using the definition of the homogeneous function and also by applying Euler's theorem.

1.2: Verify Euler's theorem for the function $\psi(x, y) = exp(x/y)$.

1.3: Discuss the degree of homogeneity of the function $v(p, T) = RT/p$ where R is a constant.

1.4: By differentiating (1.16) with respect to M^k, show that the specific partial property ψ_k is homogeneous of degree zero.

1.5: Is equation (1.18) valid for open or for closed systems?

1.6: Moist air, a mixture of dry air and water vapor, may be described by the following mixture formula

$$\psi^{0,1} = (\psi_0 M^0 + \psi_1 M^1)/(M^0 + M^1)$$

Verify that for $p = $ constant, $T = $ constant the following relations are valid:

$$\psi_0 = \psi^{0,1} - m^1 \frac{\partial \psi^{0,1}}{\partial m^1}, \qquad \psi_1 = \psi^{0,1} + (1 - m^1)\frac{\partial \psi^{0,1}}{\partial m^1}$$

1.7: The equation of state for moist air may be written in the form $pv = (R_0 m^0 + R_1 m^1)T$ where R_0 and R_1 are the gas constants for dry air and for water vapor. Use the results of Problem 1.6 to find the specific partial volumes v_0 and v_1.

1.8: (a) Show that the chain rule for Jacobians is valid:

$$J\left(\frac{u, v}{\chi, \xi}\right) = J\left(\frac{u, v}{x, y}\right) J\left(\frac{x, y}{\chi, \xi}\right)$$

where $u = u(x, y)$, $v = v(x, y)$, $x = x(\chi, \xi)$ and $y = y(\chi, \xi)$. Hint: Expand the left side and use the multiplication rule for determinants.
(b) Show that $J(u, v/x, y)J(x, y/u, v) = 1$. Interpret the results.
(c) Show that $\partial(u, y)/\partial(x, y) = (\partial u/\partial x)_y$.
(d) By using the result of (c) show that $\beta \kappa p = \alpha$ where

$$\beta = \frac{1}{p}\left(\frac{\partial p}{\partial T}\right)_v, \qquad \kappa = -\frac{1}{v}\left(\frac{\partial v}{\partial T}\right)_p, \qquad \alpha = \frac{1}{v}\left(\frac{\partial v}{\partial p}\right)_T$$

2

Budget equations

Theoretical considerations often require budget equations of certain physical quantities such as mass, momentum and energy in its various forms. The purpose of this chapter is to derive the general form of a balance or budget equation which applies to all extensive quantities and those intensive quantities derived from them. Intensive variables such as p or T cannot be balanced.

Let us consider a velocity field $\mathbf{v}(\mathbf{r}, t)^1$ in three-dimensional space. A volume $V(t)$ whose volume elements $d\tau$ are moving with the velocity $\mathbf{v}(\mathbf{r}, t)$ is called a *fluid volume*. The fluid volume may change its size and form with time since all surface elements $d\mathbf{S}$ of the imaginary surface S enclosing the volume are displaced with the velocity $\mathbf{v}(\mathbf{r}, t)$ existing at their position.

Now we envision $V(t)$ to consist of a certain number of particles also moving with $\mathbf{v}(\mathbf{r}, t)$. Obviously, at all times the particles that are located on the surface of the volume remain there because the particle velocity and the velocity of the corresponding surface element $d\mathbf{S}$ coincide. Thus, no particle can leave the volume so that the number of particles within $V(t)$ remains constant. Therefore, a fluid volume is also called a *material volume*.

In analogy to the fluid or material volume we define the *fluid* or *material surface* and the *fluid* or *material line*. As in the case of the material volume, a material surface and a material line consist of a constant number of particles.

The concept of fluid volumes may be extended to situations with different groups of particles where each group is moving with a different velocity $\mathbf{v}_k$. For our purposes we consider an atmospheric fluid volume consisting of the partial masses M^k, $k = 0, \ldots, 3$ which are displaced with the velocities $\mathbf{v}_k(\mathbf{r}, t)$, $k = 0, \ldots, 3$. Now each volume element will move with the *barycentric velocity* $\mathbf{v}(\mathbf{r}, t)$ which is

---

[1] It should be pointed out that the velocity refers to the inertial system.

defined by means of

$$\rho(\mathbf{r}, t)\mathbf{v}(\mathbf{r}, t) = \sum_{k=0}^{3} \rho^k(\mathbf{r}, t)\mathbf{v}_k(\mathbf{r}, t) \quad \text{with} \quad \rho(\mathbf{r}, t) = \sum_{k=0}^{3} \rho^k(\mathbf{r}, t) \qquad (2.1)$$

Since each surface element $d\mathbf{S}$ of the volume is also moving with the barycentric velocity $\mathbf{v}(\mathbf{r}, t)$ existing at its position, we observe a mass flux $\mathbf{J}^k$ of the partial masses M^k through $d\mathbf{S}$ which is defined by $\mathbf{J}^k \cdot d\mathbf{S}$ with

$$\mathbf{J}^k = \rho^k(\mathbf{v}_k - \mathbf{v}) \qquad (2.2)$$

Hence, $\mathbf{J}^k \cdot d\mathbf{S}$ describes the *diffusion* of the partial mass M^k through the surface element $d\mathbf{S}$ with the *diffusion velocity* $\mathbf{v}_{k,d} = \mathbf{v}_k - \mathbf{v}$. The vector $\mathbf{J}^k$ is called the *diffusion flux* of the substance k. From (2.1) and (2.2) it is easily seen that the sum of all diffusion fluxes vanishes

$$\sum_{k=0}^{3} \mathbf{J}^k = 0 \qquad (2.3)$$

We conclude that the diffusion of the partial masses M^k through the surface of the fluid volume may change the mass composition of the volume. However, due to (2.3) at all times the volume conserves the total mass.

2.1 General form of the budget equation

Usually any extended part of the atmospheric continuum will be inhomogeneous. To handle an inhomogeneous section of the atmosphere we mentally isolate a fluid volume V that is drifting with the fluid. The volume is surrounded by an imaginary surface $S = S(\mathbf{r}, t)$. Anything within S belongs to the system, the surroundings or the outside world is found exterior to S. Processes taking place at the surface S itself which contribute to changes of the system represent the interaction of the system with the surroundings, see also Figure 1.1. These processes include mass fluxes penetrating the surface, work contributions by or on the system resulting from surface forces, and heat and radiative fluxes through the surface.

We recall the definition $\Psi = \widehat{\psi} V$ where $\widehat{\psi} = \rho\psi$. In general, the density $\widehat{\psi}$ varies with time and space. The amount of Ψ contained in the volume V is then given by

$$\Psi(\mathbf{r}, t) = \int_V \widehat{\psi}(\mathbf{r}, \mathbf{r}', t)\,d\tau = \int_V \rho(\mathbf{r}, \mathbf{r}', t)\psi(\mathbf{r}, \mathbf{r}', t)\,d\tau \qquad (2.4)$$

Ψ-changes caused by the interaction of the system with the exterior are called *external changes* and will be denoted by $d_e\Psi/dt$ while *internal changes* of Ψ will be written as $d_i\Psi/dt$. The budget equation for Ψ is then given by

$$\frac{d\Psi}{dt} = \frac{d_e\Psi}{dt} + \frac{d_i\Psi}{dt} \qquad (2.5)$$

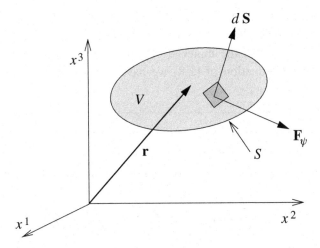

Fig. 2.1 Terms appearing in the budget equation for Ψ.

The external time change $d_e\Psi/dt$ may be expressed by an integral over the surface S bounding V,

$$\frac{d_e\Psi}{dt} = -\oint_S \mathbf{F}_\psi(\mathbf{r}', t) \cdot d\mathbf{S} = -\int_V \nabla \cdot \mathbf{F}_\psi(\mathbf{r}', t)\, d\tau \qquad (2.6)$$

where $\mathbf{F}_\psi$ is a flux vector giving the amount of Ψ per unit surface area and unit time streaming into V or out of it. The vector $d\mathbf{S}$ represents any surface area element of S and is directed toward the exterior of V as shown in Figure 2.1. The negative sign is chosen so that $d_e\Psi/dt$ is positive when $\mathbf{F}_\psi$ is directed into V. The conversion from the surface to the volume integral follows from the Gauss divergence theorem.

If Q_ψ represents the production of Ψ within V per unit volume and time, also known as *source strength*, then we may write

$$\frac{d_i\Psi}{dt} = \int_V Q_\psi\, d\tau \qquad (2.7)$$

Q_ψ is positive in case of production of Ψ (source) and negative in case of destruction (sink).

Differentiation of (2.4) with respect to time yields

$$\frac{d\Psi}{dt} = \frac{d}{dt}\left(\int_V \rho\psi\, d\tau\right) = \int_V \frac{d}{dt}(\rho\psi)\, d\tau + \int_V \rho\psi\left[\frac{1}{d\tau}\frac{d}{dt}(d\tau)\right] d\tau$$

$$= \int_V \frac{d}{dt}(\rho\psi)\, d\tau + \int_V \rho\psi\nabla\cdot\mathbf{v}\, d\tau = \int_V \frac{D}{Dt}(\rho\psi)\, d\tau \qquad (2.8)$$

where $d/dt = \partial/\partial t + \mathbf{v}\cdot\nabla$ and the product rule of differential calculus has been applied. In the third equation the relative change of the volume element

$1/d\tau[d/dt(d\tau)]$ has been replaced by the velocity divergence. This replacement is valid for fluid volume elements and is shown in the Appendix to this chapter. The time differentiation of the material or fluid volume integral is attributed to Lagrange. In the last equation of (2.8) we have introduced the *budget operator* D/Dt which is defined as[2]

$$\boxed{\frac{D}{Dt}(\rho\psi) = \frac{d}{dt}(\rho\psi) + \rho\psi\nabla\cdot\mathbf{v} = \frac{\partial}{\partial t}(\rho\psi) + \nabla\cdot(\rho\psi\mathbf{v})} \qquad (2.9)$$

Combining (2.5)–(2.9) results in the integral form of the budget equation

$$\frac{d\Psi}{dt} = \int_V \frac{D}{Dt}(\rho\psi)\,d\tau = \int_V \left[\frac{\partial}{\partial t}(\rho\psi) + \nabla\cdot(\rho\psi\mathbf{v})\right]d\tau = \int_V [-\nabla\cdot\mathbf{F}_\psi + Q_\psi]\,d\tau$$
$$(2.10)$$

This expression describes the following physical processes:

    (i) Local time change of Ψ: $\displaystyle\int_V \frac{\partial}{\partial t}(\rho\psi)\,d\tau$

    (ii) Convective flux of Ψ through S: $\displaystyle\int_V \nabla\cdot(\rho\psi\mathbf{v})\,d\tau$

    (iii) Nonconvective flux of Ψ through S: $\displaystyle -\int_V \nabla\cdot\mathbf{F}_\psi\,d\tau$

    (iv) Production or destruction of Ψ: $\displaystyle\int_V Q_\psi\,d\tau$

A particular example of a nonconvective flux is the heat conduction while phase changes of cloud water are represented by the internal source Q_ψ. Since equation (2.10) is valid for an arbitrary volume V, we immediately obtain the general differential form of the budget equation as

$$\boxed{\frac{D}{Dt}(\rho\psi) = -\nabla\cdot\mathbf{F}_\psi + Q_\psi} \qquad (2.11)$$

A very simple but extremely important example of a budget equation is the *continuity equation* describing the conservation of mass of the material volume. Identifying $\Psi = M = $ constant we have $d\Psi/dt = 0$ and $\psi = 1$ so that the differential form of equation (2.10) is now given by

$$\boxed{\frac{D\rho}{Dt} = \frac{\partial\rho}{\partial t} + \nabla\cdot(\rho\mathbf{v}) = \frac{d\rho}{dt} + \rho\nabla\cdot\mathbf{v} = 0} \qquad (2.12)$$

The continuity equation is a fundamental part of all prognostic meteorological systems.

---

[2] It is noteworthy that the budget operator is not a real differential operator since the product rule is not obeyed.

Expanding (2.9) and utilizing the continuity equation we obtain

$$\frac{D}{Dt}(\rho\psi) = \psi\left[\frac{\partial\rho}{\partial t} + \nabla\cdot(\rho\mathbf{v})\right] + \rho\left[\frac{\partial\psi}{\partial t} + \mathbf{v}\cdot\nabla\psi\right] = \rho\frac{d\psi}{dt} \qquad (2.13)$$

This identity is called the *interchange rule*. Thus, the budget equation (2.11) may also be written in the form

$$\rho\frac{d\psi}{dt} = -\nabla\cdot\mathbf{F}_\psi + Q_\psi \qquad (2.14)$$

Division of (2.5) by the constant total mass M gives

$$\frac{d\psi}{dt} = \frac{d_e\psi}{dt} + \frac{d_i\psi}{dt} \qquad (2.15)$$

Substituting this expression into (2.14) yields for the external and internal ψ-changes

$$\rho\frac{d_e\psi}{dt} = -\nabla\cdot\mathbf{F}_\psi, \qquad \rho\frac{d_i\psi}{dt} = Q_\psi \qquad (2.16)$$

It is worthwhile reconsidering (2.14) using an argument due to *Van Mieghem* (1973). Suppose we add the arbitrary vector $\mathbf{X}$ to $\mathbf{F}_\psi$ so that $\nabla\cdot\mathbf{X} = \beta$ and add β to the production term Q_ψ on the right-hand side of this equation. This mathematical operation does not change the budget equation in any way. Consequently, there is no unique definition of either $\mathbf{F}_\psi$ or Q_ψ. However, by making a proper choice of $\mathbf{F}_\psi$, the term Q_ψ will also be determined properly.

2.2 Meteorological applications to the budget equation

In this section we consider the budget equations for the concentrations m^k of the partial masses M^k, $k = 0, \ldots, 3$. Thus, we set $\Psi = M^k$ and $\psi = m^k$. Since the partial mass M^k is moving with the velocity $\mathbf{v}_k$, diffusion occurs through each surface element $d\mathbf{S}$ of the volume with the diffusion flux $\mathbf{J}^k$. For the partial masses diffusion is the only mass interaction of the volume with the surroundings, that is $\mathbf{F}_{m^k} = \mathbf{J}^k$.

Denoting internal changes of m^k by I^k we obviously have $I^0 = 0$ since the mass concentration of dry air cannot be produced or destroyed by internal processes within V. For $k = 1, 2, 3$, however, I^k describes the concentration changes due to transitions between the various phases of water (vapor, liquid and ice). Utilizing these definitions in (2.11) we obtain the *budget equation for the mass concentrations*

$$\rho \frac{dm^k}{dt} = \frac{D}{Dt}(\rho m^k) = -\nabla \cdot \mathbf{J}^k + I^k \tag{2.17}$$

Summation of this equation over all k shows that the sum over all phase transition rates must vanish

$$-\sum_{k=0}^{3} \nabla \cdot \mathbf{J}^k + \sum_{k=0}^{3} I^k = 0 \implies \sum_{k=1}^{3} I^k = 0 \tag{2.18}$$

where use has been made of (1.1) and (2.3). Without difficulty we could have obtained this result from a physical argument by observing that phase transitions cannot change the total mass of water within the fluid volume. An example may help to clarify the role of I^k. Let us consider a very simple equilibrium system consisting only of dry air, water vapor and liquid water at temperature T_1. Heating the system to temperature T_2 will cause some of the liquid water to evaporate and a new equilibrium state will be established. The decrease in liquid water is compensated by an increase of water vapor, but the total mass has not changed. The phase transition taking place is then described by $I^1 = -I^2$. Cooling the system from T_2 to T_1 will reverse the process.

To further clarify the important concept of the diffusion process dm^k/dt will be split into external and internal changes yielding for (2.17)

$$\rho \frac{dm^k}{dt} = \rho \left(\frac{d_e m^k}{dt} + \frac{d_i m^k}{dt} \right) = -\nabla \cdot \mathbf{J}^k + I^k \tag{2.19}$$

The internal change of m^k must be associated with the phase transition while the external change is identified with the divergence of the diffusion fluxes

$$\rho \frac{d_i m^k}{dt} = I^k, \qquad \rho \frac{d_e m^k}{dt} = -\nabla \cdot \mathbf{J}^k \tag{2.20}$$

Since $\psi = \psi_n m^n$ we may write

$$\begin{aligned}
\frac{d\psi}{dt} &= \frac{d}{dt}(\psi_n m^n) = m^n \frac{d\psi_n}{dt} + \psi_n \frac{dm^n}{dt} \\
&= m^n \frac{d\psi_n}{dt} + \psi_n \left(\frac{d_e m^n}{dt} + \frac{d_i m^n}{dt} \right)
\end{aligned} \tag{2.21}$$

Substitution of (2.20) into this expression yields the useful expression

$$\frac{d\psi}{dt} = m^n \frac{d\psi_n}{dt} - v\psi_n \nabla \cdot \mathbf{J}^n + v\psi_n I^n \tag{2.22}$$

2.3 Summary and additional remarks

In this chapter we have derived the general form of the budget equation. As a special case we have obtained the continuity equation describing the conservation of mass. In the budget equation we have identified various physical quantities such as the convective fluxes, the nonconvective fluxes and the production/source term. The budget equation for the mass concentrations m^k has been derived, and it has been shown that the diffusion fluxes and the phase transition fluxes add up to zero.

By utilizing the results obtained in this chapter we will now discuss in some more detail the characteristics of various types of atmospheric systems which have already been introduced in the previous chapter.

2.3.1 Closed systems

Isolated systems which do not experience exchange of mass and energy in any form are also called *adiabatic*. In the atmosphere these conditions are rarely satisfied. If condensation particles form due to rising air, some of the condensate leaves the air parcel by gravitational settling thus removing heat and mass from the system. The extreme case where all condensation particles fall out as soon as they are formed is called a *pseudoadiabatic process*. Details will be given later.

2.3.2 Systems closed with respect to mass

In an atmospheric system which is closed with respect to mass, the exchange of immaterial fluxes is permitted. Obviously, this system always contains the same mass particles so that the velocities of the partial masses on each surface element are identical, that is $v_k = v$ and $J^k = 0$, $k = 0, 1, 2, 3$.

2.3.3 Open systems

(a) A spatially fixed system surrounded by an imaginary surface is placed in the atmospheric continuum. Material substance will enter and leave the system. The system does not conserve mass.

(b) The system is moving with the velocity v_k of the partial mass M^k. Since the surface of the system is also moving with v_k, according to (2.3) the diffusion flux $J^k = 0$. In this case the system is closed with respect to the particular mass M^k while the exchange of other partial masses is possible.

(c) Each surface element of the system is moving with the barycentric velocity $v = v_n m^n$ with $v \neq v_k$. This type of open system conserves the total mass M, but it does not conserve the partial mass M^k because $J^k \neq 0$.

2.4 Appendix

Application of the Gauss divergence theorem to the velocity field $\mathbf{v}$ yields

$$\int_V \nabla \cdot \mathbf{v}(\mathbf{r}, t) d\tau = \oint_S d\mathbf{S} \cdot \mathbf{v}(\mathbf{r}, t) \tag{2.23}$$

During the time increment Δt each surface element $d\mathbf{S}(\mathbf{r}, t)$ of the fluid volume $V(t)$ is moving with the velocity $\mathbf{v}(\mathbf{r}, t)$ thereby changing the volume to $V(t + \Delta t)$. The surface elements are displaced the distance $\Delta \mathbf{r} = \mathbf{v}(\mathbf{r}, t)\Delta t$ so that the volume element of the difference volume is given by

$$\Delta V = V(t + \Delta t) - V(t) = \oint_S d\mathbf{S} \cdot \Delta \mathbf{r} \tag{2.24}$$

Using the mean value theorem for integrals, (2.23) can now be written as

$$\overline{\nabla \cdot \mathbf{v}} = \frac{1}{V} \lim_{\Delta t \to 0} \frac{1}{\Delta t} \oint_S d\mathbf{S} \cdot \Delta \mathbf{r} = \frac{1}{V} \lim_{\Delta t \to 0} \frac{\Delta V}{\Delta t} = \frac{1}{V} \frac{dV}{dt} \tag{2.25}$$

For an infinitesimally small volume $V \longrightarrow d\tau$ the term $\nabla \cdot \mathbf{v}$ is constant within $d\tau$ and we obtain

$$\boxed{\nabla \cdot \mathbf{v} = \frac{1}{d\tau} \frac{d}{dt}(d\tau)} \tag{2.26}$$

2.5 Problems

2.1: Transform the equation

$$\frac{\partial \psi}{\partial t} + \mathbf{v}_k \cdot \nabla \psi = 0$$

to the form of a budget equation. Which well-known relation results by setting $\psi = 1$?

2.2: Show that

$$\frac{D^k \rho^k}{Dt} = \frac{\partial \rho^k}{\partial t} + \nabla \cdot (\mathbf{v}_k \rho^k) = \rho \frac{dm^k}{dt} + \nabla \cdot \mathbf{J}^k$$

is a valid expression.

2.3: Transform equation (2.17) to the following expression:

$$\frac{D^k \rho^k}{Dt} = \frac{\partial \rho^k}{\partial t} + \nabla \cdot (\mathbf{v}_k \rho^k) = I^k$$

2.4: Show that it is possible to write dM^k in the form

$$dM^k = d_e M^k + \sum_{j=2}^{3} V_j^k d_i M^j \quad \text{with} \quad V_j^k = \begin{pmatrix} 0 & 0 \\ -1 & -1 \\ 1 & 0 \\ 0 & 1 \end{pmatrix} \begin{array}{l} k = 0 \\ k = 1 \\ k = 2 \\ k = 3 \end{array}$$

The term V_j^k is the so-called stoichiometric matrix.

3

The first law of thermodynamics

3.1 Introduction

Experience shows that in all thermodynamic systems the *state variables* temperature, internal energy and pressure assume a prominent role. Many experiments have been performed whose description require the introduction of temperature. This fact is expressed in the so-called *zeroth law of thermodynamics* which simply states that every thermodynamic system is associated with a variable of state T, called *temperature*. If two systems are in *thermal equilibrium*, the temperature of both systems must be identical.

Before discussing the all-important first law of thermodynamics it will be necessary to describe briefly the concepts of heat and work. If two isolated systems are brought in thermal contact by a conducting wall, heat will flow from the warmer to the colder system as long as a temperature difference exists. *Heat* is energy in transition between the systems. It is not a property of the physical system, therefore, it is not a variable of state. Thus, it makes no sense to speak of the heat of a particular system. The amount of heat that is transferred depends on the "path" for how it is added to the system. If, for example, the temperature of unit mass of air is to be increased by a certain number of degrees, it makes a significant difference if the heat is added at constant pressure or at constant volume. The amount of energy added to a system will be denoted by $đQ$ where the differential operator $đ$ expresses that $đQ$ is not an exact differential.

Energy cannot only be added to a system in terms of heat but also by doing *work* on the system by an *external force*. Vice versa, the system may also perform work on its *surroundings*. A simple example will demonstrate the principle of work as used in thermodynamics.

A gas contained in a cylinder of cross-section df exerts the pressure p on the piston which is held in equilibrium by the external force $K = p\,df$, see Figure 3.1. If the piston is pushed into the cylinder by a small distance dx, the

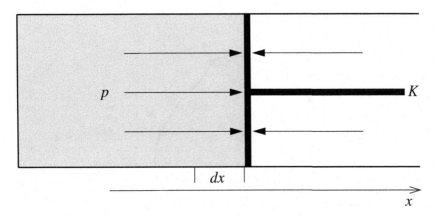

Fig. 3.1 Compression work done on a gas by a change of volume.

amount of *compression work* done on the gas is

$$đA = -K\,dx = -p\,df\,dx = -p\,dV \implies A = -\int_{V_1}^{V_2} p\,dV \qquad (3.1)$$

Since dV is negative, the work done on the system is positive and thus has the same sign as heat added to the system.

The term $đA$ is not an exact differential so that work depends on the path as shown in Figure 3.2. The fact that work is path dependent is demonstrated by considering in the (V, p)-plane the closed path from point P_1 along C_1 to point P_2 and back to point P_1 along C_2. The area enclosed between the two curves represents work. If two different curves between the two points had been chosen the area between the two curves would be different. This is expressed by the formula

$$\oint đA \neq 0 \qquad (3.2a)$$

In contrast to this, the closed line integral for a state variable Ψ is zero. If this integral extends from point P_1 to point P_2, the result depends only on the endpoints and, therefore, is independent of the path taken

$$\oint d\Psi = 0, \qquad \int_{P_1}^{P_2} d\Psi = \Psi(P_2) - \Psi(P_1) \qquad (3.2b)$$

If the motion of the piston in Figure 3.1 is produced by a finite difference between the external force and the *interior pressure force*, the piston will be accelerated and both p and V will be functions of time and the integration will become a problem of hydrodynamics. If the change of volume is done quasi-statically, p at all times

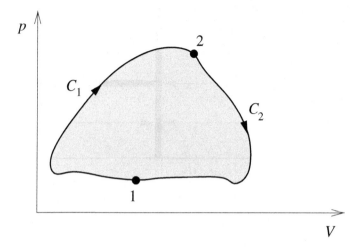

Fig. 3.2 Work diagram.

is a thermodynamic coordinate and can be expressed as a function of temperature and volume. The evaluation of the integral can be carried out if the behavior of the temperature is known so that p can be expressed in terms of V. The situation is particularly simple if an ideal gas and an isothermal expansion from V_1 to V_2 is considered, resulting in

$$A = -NR^*T \int_{V_1}^{V_2} \frac{dV}{V} = -NR^*T \ln \frac{V_2}{V_1} \tag{3.3}$$

where N is the number of moles and R^* is the universal gas constant. There are other more general cases when the integration can also be carried out directly. In some difficult situations the *work integral* must be evaluated numerically.

3.2 The internal energy and the first law of thermodynamics

To begin with, the first law of thermodynamics will be stated for a system closed with respect to mass and then followed by more general statements describing the open system. The first law of thermodynamics is conveniently formulated in terms of the following two statements:

(i) Every thermodynamic system is characterized by an extensive variable E which is called the internal energy.
(ii) Addition (subtraction) of heat $đQ$ and performance of work $đA$ on (by) the system results in the change of internal energy

$$\boxed{dE = đQ + đA} \tag{3.4}$$

The first law is a fundamental statement and cannot be reduced to something even more fundamental. For an isolated system ($đQ = 0, đA = 0$) conservation of energy is observed, i.e.

$$dE = 0, \qquad E = \text{constant} \tag{3.5}$$

The statement that the internal energy is an extensive variable of state means that E is uniquely determined by the independent variables describing the system. Some important properties of the internal energy will be summarized in a number of statements:

(i) According to (2.5), dE can be expressed by means of

$$dE = d_e E + d_i E \tag{3.6}$$

The first law of thermodynamics in the form (3.5), however, led to the conclusion that for an isolated system ($d_e E = 0$) the internal energy is conserved so that $d_i E$ must be zero. In other words, internal energy cannot be produced within the interior of the system.
(ii) Energy can be transferred between systems in terms of heat and/or work.
(iii) Energy in a particular form can be transformed into other forms of energy. This will be illustrated by a simple example. Suppose that a cylinder containing a gas is perfectly isolated against heat transfer. If work is done on the gas by compression this will result in a temperature increase of the system and, therefore, in an increase of the internal energy. If the cylinder is then brought into thermal contact with a colder body, energy is removed from the system in the form of heat.

The first law of thermodynamics does not prohibit the complete transformation of energy from one form to another. The second law of thermodynamics, however, introduces some restrictions.

Since the internal energy is an extensive quantity it is homogeneous of degree one so that Euler's theorem (1.14) applies,

$$E(p, T, M^k) = \left(\frac{\partial E}{\partial M^n} \right)_{p,T} M^n = e_n M^n \quad \text{or intensively} \quad e = e_n m^n \tag{3.7}$$

All statements derived for Ψ or ψ are also valid for E or e.

So far the first law referred to systems which were not permitted to exchange mass with the surroundings. If this restriction is removed, it becomes problematical to use the first law in the extensive form $dE = đQ + đA$ because of the definition of work. This can be illustrated by an example nicely described by Kluge and Neugebauer (1976). A box containing a gas of temperature T and pressure p will be subdivided by a movable wall into two equally large compartments as shown

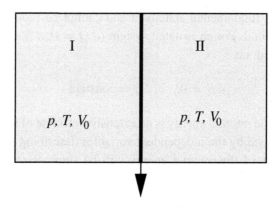

Fig. 3.3 Doubling of the volume without the performance of work.

in Figure 3.3. The thermodynamic system to be considered is compartment I having internal energy E and volume V_0. The separating wall will be removed (in the ideal case no work is done) so that the mass confined within II is added to I thus doubling the internal energy and the volume V_0 without adding heat and doing work on system I. Therefore, the differential of work $d\!\!\!/A = -p\,dV$ is not a suitable definition for a system which is open with respect to mass. The situation is different if we consider the change of internal energy per unit mass since it is independent of the addition of mass. The change of internal energy depends only on the heat per unit mass $d\!\!\!/q$ added to the system and the work done on unit mass $d\!\!\!/a = -p\,dv$. Equally well, we could have referred to the change of the internal energy per mole. This example teaches that the first law in the form

$$de = d\!\!\!/q + d\!\!\!/a \tag{3.8}$$

is valid for systems open as well as closed with respect to mass. In meteorological analyses we usually refer to unit mass since very seldom the mass of an atmospheric system is known.

The next step in the derivation of the prognostic equations of the atmosphere is the development of expressions for the time change of heat and work. To make things simple, we shall first consider a system closed with respect to mass and then generalize.

3.2.1 The atmospheric system without diffusion

To eliminate diffusion fluxes we must set $\mathbf{v}_k = \mathbf{v}$ in (2.2). We will assume that heat is added to the system bounded by an imaginary surface surrounding the volume V by means of heat conduction $\mathbf{J}$ and by *radiative transfer* $\mathbf{F}_R$. Thus we may write

$$\frac{dQ}{dt} = -\oint_S (\mathbf{J} + \mathbf{F}_R) \cdot d\mathbf{S} = -\int_V \nabla \cdot (\mathbf{J} + \mathbf{F}_R) \, d\tau \tag{3.9}$$

We assert that this energy will solely be used to change the internal energy of the system. Mechanical energies of the system will not be affected.

Next, we consider the *surface forces* acting on a surface element. These result from the *pressure force* $-p\,d\mathbf{S}$ acting in the direction opposite to $d\mathbf{S}$ and from the normal and tangential *viscous forces* $\mathbb{F} \cdot d\mathbf{S}$ which depend on the state of motion of the medium. The quantity $\mathbb{F}$ is known as the *viscous stress tensor* which is assumed to be symmetric. At this point the tensor $\mathbb{F}$ is introduced only formally but will be discussed in more detail in later chapters. The surface forces are then given by

$$-pd\mathbf{S} + \mathbb{F} \cdot d\mathbf{S} = -(p\mathbb{E} - \mathbb{F}) \cdot d\mathbf{S} \tag{3.10}$$

where $\mathbb{E}$ is the unit dyadic.[1]

Work is defined as the scalar product of a force and the displacement vector $d\mathbf{r}$ so that work per unit time due to these forces is given by

$$\frac{dA}{dt} = -\oint_S p\mathbf{v} \cdot d\mathbf{S} + \oint_S \mathbb{F} \cdot \mathbf{v} \cdot d\mathbf{S} = -\int_V \nabla \cdot (p\mathbf{v}) \, d\tau + \int_V \nabla \cdot (\mathbb{F} \cdot \mathbf{v}) \, d\tau \tag{3.11}$$

The expansion of the two divergence expressions results in two work terms

$$\frac{dA_E}{dt} = -\int_V p\nabla \cdot \mathbf{v} \, d\tau + \int_V \mathbb{F} \cdot\cdot \nabla\mathbf{v} \, d\tau \tag{3.12}$$

and

$$\frac{dA_M}{dt} = -\int_V \mathbf{v} \cdot \nabla p \, d\tau + \int_V \mathbf{v} \cdot \nabla \cdot \mathbb{F} \, d\tau \tag{3.13}$$

Energy budget considerations (see e.g. Chapter 1 of *DA*) show that only (3.12) contributes to a change of the internal energy whereas the part dA_M/dt contributes to the mechanical energy of the system. Therefore, we exclude dA_M/dt from further considerations concerning the first law. In case of systems homogeneous with repect to p and $\mathbb{F}$ (i.e. $\nabla p = 0$, $\nabla \cdot \mathbb{F} = 0$) one obtains $dA_M/dt = 0$.

The only contribution to the internal energy of the fluid volume system without the inclusion of diffusion is, therefore,

$$\frac{dE}{dt} = \frac{d}{dt}\int_V \rho e\, d\tau = \int_V \frac{D}{Dt}(\rho e) \, d\tau = \int_V \rho \frac{de}{dt} \, d\tau$$
$$= -\int_V [\nabla \cdot (\mathbf{J} + \mathbf{F}_R) + p\nabla \cdot \mathbf{v} - \mathbb{F} \cdot\cdot \nabla\mathbf{v}] \, d\tau \tag{3.14}$$

---

[1] At this point the reader may consult Appendix A (Section 3.8) of this chapter where dyadics in the Cartesian system are briefly introduced.

Since (3.14) is valid for an arbitrary volume, the differential form follows immediately,

$$\frac{de}{dt} = -v\nabla \cdot (\mathbf{J} + \mathbf{F}_{\mathrm{R}}) - vp\nabla \cdot \mathbf{v} + v\mathbb{F} \cdot\cdot \nabla\mathbf{v} \tag{3.15a}$$

By observing that $v = 1/\rho$ we find from the continuity equation (2.12) that the term $vp\nabla \cdot \mathbf{v}$ represents *compression work* since $v\nabla \cdot \mathbf{v} = dv/dt$. Equation (3.15a) is the first law per unit mass and time. This can, of course, be written as

$$\frac{de}{dt} = \frac{đq}{dt} + \frac{đa_{\mathrm{E}}}{dt} \quad \text{with}$$

$$\frac{đq}{dt} = -v\nabla \cdot (\mathbf{J} + \mathbf{F}_{\mathrm{R}}), \qquad \frac{đa_{\mathrm{E}}}{dt} = -vp\nabla \cdot \mathbf{v} + v\mathbb{F} \cdot\cdot \nabla\mathbf{v} \tag{3.15b}$$

In this expression the general form for de/dt is valid for all systems whereas here the specifications for $đq/dt$, $đa_{\mathrm{E}}/dt$ refer to a system without diffusion.

3.2.2 The atmospheric system with diffusion

In contrast to a system without diffusion, where $\mathbf{v}_k = \mathbf{v}$, we must now admit that $\mathbf{v}_k \neq \mathbf{v}$ so that $\mathbf{J}^k \neq 0$. This leads to the following modifications:

(i) To the immaterial fluxes $\mathbf{J}$ and $\mathbb{F}$, material components must be added. We think of these components to be included in $\mathbf{J}$ and $\mathbb{F}$ by formally replacing $\mathbf{J}$ by $\mathbf{J}^h$ and $\mathbb{F}$ by $\mathbb{J}$.
(ii) Additional contributions to $đa_{\mathrm{E}}/dt$ must be included.

The task ahead is to reformulate the *budget equation for the internal energy* by now admitting the effect of diffusion as part of $đa_{\mathrm{E}}/dt$. For this open system we use the first law and other equations in their intensive form. We shall proceed by first formulating the balance equation for the *total energy* ϵ which is a conservative quantity so that the source term Q_ϵ vanishes. Since we are dealing with the absolute system of motion, ϵ is given by the sum of the specific internal energy e, the total specific kinetic energy k and the *specific potential energy* ϕ_{a} which is also known as the *attractional potential*, $\epsilon = e + k + \phi_{\mathrm{a}}$. The total kinetic energy consists of the *kinetic energy of the barycentric motion k_{b}* and the *kinetic energy of the diffusion motion k_{d}*

$$k = \frac{1}{2}\sum_{k=0}^{3} m^k \mathbf{v}_k \cdot \mathbf{v}_k = \frac{1}{2}\sum_{k=0}^{3} m^k (\mathbf{v}_{k,\mathrm{d}} + \mathbf{v}) \cdot (\mathbf{v}_{k,\mathrm{d}} + \mathbf{v})$$

$$= \frac{1}{2}m^n \mathbf{v}_{n,\mathrm{d}}^2 + \frac{\mathbf{v}^2}{2} = k_{\mathrm{d}} + k_{\mathrm{b}} \quad \text{with} \tag{3.16}$$

$$k_{\mathrm{d}} = \frac{1}{2}m^n \mathbf{v}_{n,\mathrm{d}}^2, \qquad k_{\mathrm{b}} = \frac{\mathbf{v}^2}{2}$$

In this expression use has been made of $\sum_{k=0}^{3} m^k \mathbf{v}_{k,\mathrm{d}} = 0$. Thus, the total energy may be written as

$$\epsilon = e + k_{\mathrm{b}} + k_{\mathrm{d}} + \phi_{\mathrm{a}} \tag{3.17}$$

and the corresponding conservation equation, see (2.11), as

$$\frac{D}{Dt}(\rho\epsilon) + \nabla \cdot \mathbf{F}_{\epsilon} = 0 \tag{3.18}$$

Denoting the *mechanical energy* by $\epsilon_{\mathrm{M}} = k_{\mathrm{b}} + \phi_{\mathrm{a}}$, then the budget equation for ϵ_{M} assumes the form

$$\frac{D}{Dt}(\rho\epsilon_{\mathrm{M}}) + \nabla \cdot \mathbf{F}_{\epsilon_{\mathrm{M}}} = Q_{\epsilon_{\mathrm{M}}} \tag{3.19}$$

where now the source term differs from zero. Without proof we accept the budget equation for ϵ_{M} to identify $\mathbf{F}_{\epsilon_{\mathrm{M}}}$ and $Q_{\epsilon_{\mathrm{M}}}$

$$\frac{D}{Dt}\left[\rho\left(k_{\mathrm{b}} + \phi_{\mathrm{a}}\right)\right] + \nabla \cdot \left[\mathbf{v} \cdot (p\mathbb{E} - \mathbb{J})\right] = (p\mathbb{E} - \mathbb{J}) \cdots \nabla\mathbf{v} \tag{3.20}$$

A detailed derivation of this equation may be found in *DA*. The *work* done per unit time by surface forces is given by $\mathbf{F}_{\epsilon_{\mathrm{M}}} = \mathbf{v} \cdot (p\mathbb{E} - \mathbb{J})$ and the internal production of mechanical energy by $Q_{\epsilon_{\mathrm{M}}} = (p\mathbb{E} - \mathbb{J}) \cdots \nabla\mathbf{v}$.

In order to derive a budget equation for e, we subtract (3.19) from (3.18) using (3.17) to obtain

$$
\begin{aligned}
\text{(a)} \quad & \frac{D}{Dt}(\rho e) + \frac{D}{Dt}(\rho k_{\mathrm{d}}) + \nabla \cdot (\mathbf{F}_{\epsilon} - \mathbf{F}_{\epsilon_{\mathrm{M}}}) = -Q_{\epsilon_{\mathrm{M}}} \quad \text{or} \\
\text{(b)} \quad & \frac{de}{dt} = -v\nabla \cdot (\mathbf{F}_{\epsilon} - \mathbf{F}_{\epsilon_{\mathrm{M}}}) - vQ_{\epsilon_{\mathrm{M}}} - \frac{dk_{\mathrm{d}}}{dt}
\end{aligned} \tag{3.21}
$$

Comparison with (3.15b) and recalling that now $\mathbf{J}$, $\mathbb{F}$ must be replaced by $\mathbf{J}^h$, $\mathbb{J}$, yields

$$\frac{dq}{dt} = -v\nabla \cdot (\mathbf{F}_{\epsilon} - \mathbf{F}_{\epsilon_{\mathrm{M}}}) = -v\nabla \cdot (\mathbf{J}^h + \mathbf{F}_{\mathrm{R}}) \tag{3.22}$$

and

$$\frac{da_{\mathrm{E}}}{dt} = -vQ_{\epsilon_{\mathrm{M}}} - \frac{dk_{\mathrm{d}}}{dt} = -v(p\mathbb{E} - \mathbb{J}) \cdots \nabla\mathbf{v} - \frac{dk_{\mathrm{d}}}{dt} \tag{3.23}$$

Adding (3.22) and (3.23) gives the first law of thermodynamics in the customary form (3.15), but now dq/dt and da_{E}/dt have an extended meaning.

In order to evaluate (3.21a) we must derive an expression for $D(\rho k_{\mathrm{d}})/Dt$ by simply substituting k_{d} from (3.16) and then using the budget equation (2.19) for concentrations. Application of the definition (2.2) for the diffusion flux results in

$$\frac{D}{Dt}(\rho k_{\mathrm{d}}) = \frac{\rho}{2}\frac{d}{dt}\left(m^n \mathbf{v}_{n,\mathrm{d}}^2\right) = \mathbf{J}^n \cdot \frac{d\mathbf{v}_{n,\mathrm{d}}}{dt} + \frac{\mathbf{v}_{n,\mathrm{d}}^2}{2}\left(-\nabla \cdot \mathbf{J}^n + I^n\right) \tag{3.24}$$

or in the more suitable form

$$\frac{D}{Dt}(\rho k_{\mathrm{d}}) = -\nabla \cdot \left(\mathbf{J}^n \frac{\mathbf{v}_{n,\mathrm{d}}^2}{2} \right) + I^n \frac{\mathbf{v}_{n,\mathrm{d}}^2}{2} + \mathbf{J}^n \cdot \left(\frac{d\mathbf{v}_{n,\mathrm{d}}}{dt} + \nabla \frac{\mathbf{v}_{n,\mathrm{d}}^2}{2} \right) \quad (3.25)$$

With reference to (3.22) and (3.23), we finally obtain two equivalent expressions for the budget of the internal energy

$$\frac{D}{Dt}(\rho e) + \nabla \cdot (\mathbf{J}^h + \mathbf{F}_{\mathrm{R}}) = -(p\mathbb{E} - \mathbb{J}) \cdot\cdot \nabla \mathbf{v} - \rho \frac{dk_{\mathrm{d}}}{dt} \quad (3.26)$$

or

$$\frac{D}{Dt}(\rho e) + \nabla \cdot \left(\mathbf{J}^h + \mathbf{F}_{\mathrm{R}} - \mathbf{J}^n \frac{\mathbf{v}_{n,\mathrm{d}}^2}{2} \right)$$
$$= (-p\mathbb{E} - \mathbb{J}) \cdot\cdot \nabla \mathbf{v} - I^n \frac{\mathbf{v}_{n,\mathrm{d}}^2}{2} - \mathbf{J}^n \cdot \left(\frac{d\mathbf{v}_{n,\mathrm{d}}}{dt} + \nabla \frac{\mathbf{v}_{n,\mathrm{d}}^2}{2} \right) \quad (3.27)$$

In order to avoid problems associated with the mathematical treatment of the diffusion terms in (3.27), some authors extend the definition of the internal energy by including the kinetic energy of diffusion, i.e.

$$e_k^* = e_k + \frac{\mathbf{v}_{k,\mathrm{d}}^2}{2} \quad (3.28)$$

Multiplying this equation by m^k and summing over all k results in

$$e^* = \left(e_n + \frac{\mathbf{v}_{n,\mathrm{d}}^2}{2} \right) m^n = e + k_{\mathrm{d}} \quad (3.29)$$

so that (3.26) can be written as

$$\frac{D}{Dt}(\rho e^*) + \nabla \cdot (\mathbf{J}^h + \mathbf{F}_{\mathrm{R}}) = -(p\mathbb{E} - \mathbb{J}) \cdot\cdot \nabla \mathbf{v} \quad (3.30)$$

This latter equation now has the same mathematical structure as (3.15a) which represents the diffusionless system. We will not use this form of the first law but approach the problem in a simplified way. As pointed out before, the expression $vp\mathbb{E} \cdot\cdot \nabla \mathbf{v} = vp\nabla \cdot \mathbf{v}$ can be replaced by $p \, dv/dt$.

In order to obtain a simplified form of the first law, which is still amply accurate for most applications, we rewrite at first $đa_{\mathrm{E}}/dt$ in (3.23) as

$$\frac{đa_{\mathrm{E}}}{dt} = -p\frac{dv}{dt} + v\mathbb{J} \cdot\cdot \nabla \mathbf{v} - \frac{dk_{\mathrm{d}}}{dt} = -p\frac{dv}{dt} + \frac{đw}{dt} \quad (3.31)$$

where formally $d w/dt$ has been introduced for the last two terms on the right-hand side of this expression. For many situations $dk_d/dt \approx 0$ although k_d itself is not necessarily small but nearly constant in time. This would be the case, for example, if precipitation particles move with almost no acceleration. If this assumption is valid, then we approximate viscous effects by

$$\frac{d w}{dt} \approx v \mathbb{J} \cdots \nabla \mathbf{v} \tag{3.32}$$

and the first law can be written as

$$\boxed{\frac{de}{dt} + p\frac{dv}{dt} = \frac{d q}{dt} + \frac{d w}{dt} = -v\nabla \cdot (\mathbf{J}^h + \mathbf{F_R}) + v\mathbb{J} \cdots \nabla \mathbf{v}} \tag{3.33}$$

The radiative net flux $\mathbf{F_R}$ will be obtained by solving the radiative transfer equation. The mathematical forms of the heat flux vector $\mathbf{J}^h$ and the viscous flux $\mathbb{J}$ will be derived later with the help of the second law of thermodynamics.

3.3 The enthalpy

In this section the prognostic equation for the internal energy will be reformulated so that finally the temperature T will appear as the prognostic variable which is really the quantity needed in atmospheric problems. The transformation from e to T is a somewhat lengthy process, so we proceed stepwise. At the same time we will obtain some useful expressions needed later.

In atmospheric problems it is always convenient to involve temperature and pressure since these are the coordinates which are normally observed. For this reason we eliminate dv in (3.33) by applying the so-called Legendre transformation,

$$de + pdv = d(e + pv) - vdp \tag{3.34}$$

The new combination of state variables is given the name *enthalpy*

$$\boxed{h = e + pv \quad \text{or} \quad H = E + pV} \tag{3.35}$$

which is stated here in terms of intensive as well as extensive coordinates. Since (e, p, v) and (E, p, V) are variables of state, the enthalpy is a state variable also. Every function of state variables is a variable of state also. This results in the new form of the first law

$$\boxed{\frac{dh}{dt} - v\frac{dp}{dt} = \frac{d q}{dt} + \frac{d w}{dt} = -v\nabla \cdot (\mathbf{J}^h + \mathbf{F_R}) + v\mathbb{J} \cdots \nabla \mathbf{v}} \tag{3.36}$$

Equations (3.33) and (3.36) show that de is the energy $đ(q + w)$ added to a system by an *isochoric process* ($dv = 0$) while dh refers to energy added *isobarically* ($dp = 0$). The concept of enthalpy will be very useful in our further discussions. By identifying $\Psi = H$ and $\psi = h$, all former statements derived for Ψ and ψ now also apply to enthalpy. Since Ψ is homogeneous of degree one in mass coordinates, Euler's theorem also holds for enthalpy

$$H = \left(\frac{\partial H}{\partial M^n}\right)_{p,T} M^n = h_n M^n \quad \text{or} \quad h = h_n m^n = (e_n + pv_n) m^n \qquad (3.37)$$

Thus we define

$$h_k = e_k + pv_k \qquad (3.38)$$

We shall now consider enthalpy as function of the usual variables

$$h = h(p, T, m^k) \qquad (3.39)$$

Application of (1.18), (1.20) and (1.21) gives immediately

$$dh = \left(\frac{\partial h}{\partial T}\right)_{p,m^k} dT + \left(\frac{\partial h}{\partial p}\right)_{T,m^k} dp + h_n dm^n \quad \text{with}$$

$$h_n dm^n = \sum_{k=0}^{3} (h_k - h_1) dm^k, \qquad \left(\frac{\partial h}{\partial m^k}\right)_{p,T} = h_k - h_1, \qquad k \neq 1 \qquad (3.40)$$

By writing the first law in the form $dh - vdp = đ(q + w)$ we now obtain

$$dh - vdp = \left(\frac{\partial h}{\partial T}\right)_{p,m^k} dT + \left[\left(\frac{\partial h}{\partial p}\right)_{T,m^k} - v\right] dp + \sum_{k=0}^{3} (h_k - h_1) dm^k \qquad (3.41)$$

$$= đ(q + w)$$

To facilitate the physical interpretation of (3.41), dm^k will be split into external and internal parts yielding

$$\left(\frac{\partial h}{\partial T}\right)_{p,m^k} dT + \left[\left(\frac{\partial h}{\partial p}\right)_{T,m^k} - v\right] dp + (h_2 - h_1) d_i m^2 + (h_3 - h_1) d_i m^3$$

$$= đ(q + w) - h_n d_e m^n \qquad (3.42)$$

where $d_i m^0 = 0$ has been utilized.

The coefficients of the various differentials in (3.42) admit a very meaningful physical interpretation. If we simply refer to $đ(q + w)$ as the amount of heat applied to unit mass, we obtain the following important quantities:

- *specific heat at constant pressure*, i.e. the amount of heat required to change the temperature by one degree, $dp = 0$, $d_i m^2 = 0$, $d_i m^3 = 0$, $d_e m^k = 0$ for all k.

$$\frac{đ}{dT}(q + w) = \left(\frac{\partial h}{\partial T}\right)_{p,m^k} = c_p \tag{3.43a}$$

- *specific heat of tension*, i.e. the amount of heat required to increase the pressure by one unit, $dT = 0$, $d_i m^2 = 0$, $d_i m^3 = 0$, $d_e m^k = 0$ for all k.

$$\frac{đ}{dp}(q + w) = \left(\frac{\partial h}{\partial p}\right)_{T,m^k} - v = \gamma \tag{3.43b}$$

- *latent heat of vaporization*, i.e. the heat required to evaporate the unit mass of liquid water, $dT = 0$, $dp = 0$, $d_i m^3 = 0$, $d_e m^k = 0$ for all k.

$$\frac{đ}{d_i m^2}(q + w) = \left(\frac{\partial h}{\partial m^2}\right)_{p,T} = h_2 - h_1 = -l_{21} \tag{3.43c}$$

- *latent heat of sublimation*, i.e. the amount of heat required to change the unit mass of ice to vapor, $dT = 0$, $dp = 0$, $d_i m^2 = 0$, $d_e m^k = 0$ for all k.

$$\frac{đ}{d_i m^3}(q + w) = \left(\frac{\partial h}{\partial m^3}\right)_{p,T} = h_3 - h_1 = -l_{31} \tag{3.43d}$$

In order to obtain a suitable expression for the latent heat of melting l_{32}, we again consider the system with $d_e m^k = 0$. The only change assumed to take place is the conversion of ice to liquid water or vice versa so that $d_i m^2 = -d_i m^3$. We then obtain l_{32}

- *latent heat of melting*

$$\left(\frac{\partial h}{\partial m^3}\right)_{p,T} - \left(\frac{\partial h}{\partial m^2}\right)_{p,T} = h_3 - h_2 = -l_{32} \tag{3.43e}$$

No special symbol is used for terms involving m^0.

3.4 Clausius–Kirchhoff relations

In this section we will make use of the fact that the enthalpy h is a variable of state so that dh is an exact differential [see Appendix B (Section 3.9) of this chapter]. Therefore,

$$\boxed{\frac{\partial^2 h}{\partial x^i \partial x^j} = \frac{\partial^2 h}{\partial x^j \partial x^i}, \qquad x^i = p, T, m^k} \tag{3.44}$$

From this identity and the definitions listed in (3.43) a number of useful expressions follow which are known as the Clausius–Kirchhoff relations.

$$\frac{\partial^2 h}{\partial p \partial T} = \frac{\partial^2 h}{\partial T \partial p} \implies \left(\frac{\partial c_p}{\partial p}\right)_{T,m^k} = \frac{\partial}{\partial T}(\gamma + v)_{p,m^k}$$

$$\frac{\partial^2 h}{\partial m^2 \partial T} = \frac{\partial^2 h}{\partial T \partial m^2} \implies \left(\frac{\partial c_p}{\partial m^2}\right)_{p,T} = -\left(\frac{\partial l_{21}}{\partial T}\right)_{p,m^k} = c_{p,2} - c_{p,1}$$

$$\frac{\partial^2 h}{\partial m^3 \partial T} = \frac{\partial^2 h}{\partial T \partial m^3} \implies \left(\frac{\partial c_p}{\partial m^3}\right)_{p,T} = -\left(\frac{\partial l_{31}}{\partial T}\right)_{p,m^k} = c_{p,3} - c_{p,1}$$

$$\frac{\partial^2 h}{\partial m^2 \partial p} = \frac{\partial^2 h}{\partial p \partial m^2} \implies \left(\frac{\partial(\gamma + v)}{\partial m^2}\right)_{p,T} = -\left(\frac{\partial l_{21}}{\partial p}\right)_{T,m^k} = (\gamma + v)_2 - (\gamma + v)_1$$

$$\frac{\partial^2 h}{\partial m^3 \partial p} = \frac{\partial^2 h}{\partial p \partial m^3} \implies \left(\frac{\partial(\gamma + v)}{\partial m^3}\right)_{p,T} = -\left(\frac{\partial l_{31}}{\partial p}\right)_{T,m^k} = (\gamma + v)_3 - (\gamma + v)_1$$

$$(3.45)$$

The last four identities have been obtained with the help of (1.21). For convenience, the variables to be kept constant in the mixed partial derivatives have been omitted.

We shall conclude this section by deriving some additional useful relations which will help us to appreciate some mathematical operations. Recall that the derivation of Euler's theorem was based on the assumption that p, T and M^k are independent variables. We now differentiate (1.14) for $l = 1$ with respect to T,

$$\left(\frac{\partial \Psi}{\partial T}\right)_{p,M^k} = \left[\frac{\partial}{\partial M^n}\left(\frac{\partial \Psi}{\partial T}\right)_{p,M^k}\right]_{p,T} M^n \tag{3.46}$$

Setting $(\partial \Psi / \partial T)_{p,M^k} = A$ we find

$$A = \left(\frac{\partial A}{\partial M^n}\right)_{p,T} M^n \tag{3.47}$$

This is again a homogeneous function of degree one so that A is an extensive variable. Differentiation with respect to p gives an analogous result. In other words, the differentiation of an extensive variable with respect to p or T gives another extensive variable. Particularly, c_p and γ are specific values of extensive variables so that

$$c_p = \left(\frac{\partial h}{\partial T}\right)_{p,m^k} = \left(\frac{\partial h_n}{\partial T}\right)_{p,m^k} m^n = c_{p,n} m^n \quad \text{with} \quad c_{p,k} = \left(\frac{\partial h_k}{\partial T}\right)_{p,m^k} \tag{3.48a}$$

Similarly

$$\gamma = \gamma_n m^n \quad \text{with} \quad \gamma_k = \left(\frac{\partial h_k}{\partial p}\right)_{T,m^k} - v_k \tag{3.48b}$$

Moreover, according to (1.21) we find

$$\left(\frac{\partial c_p}{\partial m^k}\right)_{p,T} = c_{p,k} - c_{p,1} \quad \text{and} \quad \left(\frac{\partial \gamma}{\partial m^k}\right)_{p,T} = \gamma_k - \gamma_1 \qquad k \neq 1 \qquad (3.48c)$$

which has already been stated in (3.45).

3.5 The prognostic equation for temperature involving pressure

The only work that remains to be done in the derivation of the prognostic equation for the temperature is to substitute the coefficients summarized in (3.43) into (3.42). The first law is then converted into a prognostic equation for T by dividing all differentials by dt.

Recalling (2.20), (3.32) and (3.33) we immediately obtain from (3.42)

$$c_p \frac{dT}{dt} + \gamma \frac{dp}{dt} - vl_{21}I^2 - vl_{31}I^3 - vh_n\nabla\cdot\mathbf{J}^n + v\nabla\cdot(\mathbf{J}^h + \mathbf{F}_R) - v\mathbb{J}\cdots\nabla\mathbf{V} = 0$$

$$(3.49)$$

Two divergence terms will be combined yielding

$$-h_n\nabla\cdot\mathbf{J}^n + \nabla\cdot\mathbf{J}^h = \nabla\cdot\mathbf{J}_s^h + \mathbf{J}^n\cdot\nabla h_n \qquad (3.50)$$

where $\mathbf{J}_s^h$ is the *sensible enthalpy flux* as defined by

$$\mathbf{J}_s^h = \mathbf{J}^h - \mathbf{J}^n h_n \qquad (3.51)$$

The sensible enthalpy flux differs from $\mathbf{J}^h$ by the amount $-\mathbf{J}^n h_n$ which is known as the *diffusive enthalpy flux*. We shall see later that $\mathbf{J}_s^h$, in first approximation, is the *heat conduction flux*. Substituting (3.50) into (3.49) finally results in the prognostic equation for the temperature

$$\boxed{c_p\rho\frac{dT}{dt} = -\gamma\rho\frac{dp}{dt} + l_{21}I^2 + l_{31}I^3 - \nabla\cdot(\mathbf{J}_s^h + \mathbf{F}_R) - \mathbf{J}^n\cdot\nabla h_n + \mathbb{J}\cdots\nabla\mathbf{v}}$$

$$(3.52)$$

This equation is also called the *heat equation in terms of p*. In the following section an analogous heat equation will be derived where v appears instead of p.

Equation (3.52) is a very complete form of the heat equation. It should be noted that it is not a budget or balance equation for the temperature since intensive quantities such as temperature T and pressure p cannot be balanced. As stated before, only those intensive quantities that are derived from extensive quantities can be balanced. Thus (3.52) is a budget equation for the specific enthalpy written as a prognostic equation for the atmospheric temperature. Moreover, we should keep in mind that the heat equation is only a part of the meteorological system of equations including the continuity equation and the prognostic equations for the velocity components. Some diagnostic relations such as the ideal gas law complete the prognostic system.

The quantities I^2 and I^3 symbolize the phase transition rates and represent mass changes of the water substance per unit volume and time. Multiplication of I^2 and I^3 with the latent heat l_{21} or l_{31} gives the time change of energy per unit volume. The divergence term involves the sensible heat flux and the radiative flux integrated over the entire solar and infrared spectrum. The methods required to find the radiative flux will not be discussed in this book. The terms containing the diffusion fluxes $\mathbf{J}^k$ should not be confused with the diffusive enthalpy flux $h_n \mathbf{J}^n$. The final term involving the double scalar product between the stress tensor and the gradient of the velocity represents frictional effects or *energy dissipation*.

The reader should not concern himself with the formal evaluation of the various fluxes appearing in (3.52). This subject will be taken up in some detail later. Reference to (2.3) shows that the sum of the diffusion fluxes equals zero. If all of these fluxes were calculated independently, they would not sum precisely to zero (due to small numerical errors). To reduce the calculation work and to make sure that the diffusion fluxes sum precisely to zero, it is customary to eliminate one of these, say $\mathbf{J}^0$.

The definition of all h_k or e_k as well as other potential functions always involves an indetermined value $h_k(0)$ or $e_k(0)$ at a reference temperature T_0 and pressure p_0 so that these terms cannot not be uniquely evaluated. It will be observed that (3.52) involves only the gradient ∇h_k so that the indetermined reference enthalpies do not appear. This is a good reason for the introduction of $\mathbf{J}^h_s$. Therefore, (3.52) is given preference over (3.49).

Some final comments on the prognostic equation of temperature will conclude this section. As will be seen later, for atmospheric systems $\gamma \approx -v$ so that the two time derivatives can be combined to a single expression involving the potential temperature. Otherwise the generalized velocity $dp/dt = \omega$ can be obtained from the dynamics part of the prognostic system.

3.6 The prognostic equation for temperature involving v

As already mentioned in the previous section, it is also possible to derive the prognostic temperature equation involving v. The starting point in this derivation is the first law of thermodynamics in the coordinates (v, T)

$$de + p\,dv = đ(q + w) \tag{3.53}$$

We consider the internal energy in the coordinates $e(v, T, m^0, m^2, m^3)$. Expansion of e then results in

$$
\left(\frac{\partial e}{\partial T}\right)_{v,m^k} dT + \left[\left(\frac{\partial e}{\partial v}\right)_{T,m^k} + p\right] dv + \left(\frac{\partial e}{\partial m^2}\right)_{v,T} d_i m^2 + \left(\frac{\partial e}{\partial m^3}\right)_{v,T} d_i m^3
$$

$$
= \dt(q + w) - \sum_{k=0}^{3} \left(1 - \delta_1^k\right) \left(\frac{\partial e}{\partial m^k}\right)_{v,T} d_e m^k
$$

$$(3.54)$$

where δ_1^k is the Kronecker delta symbol. dm^k has been split in internal and external changes. Note that $(\partial e/\partial m^k)_{v,T} \neq e_k - e_1$ since in the partial differentiations (v, T) are held constant and not (p, T). Equation (3.54) is used to define the following specific quantities:

- *specific heat at constant volume,* $dv = 0, d_i m^2 = 0, d_i m^3 = 0, d_e m^k = 0$ for all k.

$$
\frac{\dt}{dT}(q + w) = \left(\frac{\partial e}{\partial T}\right)_{v,m^k} = c_v \tag{3.55a}
$$

- *specific heat of extension at constant volume,* $dT = 0, d_i m^2 = 0, d_i m^3 = 0, d_e m^k = 0$ for all k.

$$
\frac{\dt}{dv}(q + w) = \left(\frac{\partial e}{\partial v}\right)_{T,m^k} + p = \delta \tag{3.55b}
$$

- *latent heat of vaporization at constant volume,* $dT = 0, dv = 0, d_i m^3 = 0, d_e m^k = 0$ for all k.

$$
-\left(\frac{\partial e}{\partial m^2}\right)_{v,T} = l_{v,21} \tag{3.55c}
$$

- *latent heat of sublimation at constant volume,* $dT = 0, dv = 0, d_i m^2 = 0, d_e m^k = 0$ for all k.

$$
-\left(\frac{\partial e}{\partial m^3}\right)_{v,T} = l_{v,31} \tag{3.55d}
$$

- *latent heat of melting at constant volume*

$$
l_{v,32} = l_{v,31} - l_{v,21} \tag{3.55e}
$$

In contrast to c_p and γ the two quantities c_v and δ are not extensive variables so that $c_v \neq c_{v,n} m^n$ and $\delta \neq \delta_n m^n$. This is due to the fact that in the expressions in (3.55) v instead of p is taken as independent variable so that (3.46) and (3.47) do not apply. However, later it will be shown that the internal energy as well as the enthalpy of ideal gases depend only on T so that in this case $c_v = c_{v,n} m^n$ while

$\delta = p$ and $\gamma = -v$ according to (3.55b) and (3.43b). Substituting the coefficients listed in (3.55) into (3.54) yields the *heat equation in terms of v.*

$$
\boxed{
\begin{aligned}
c_v \frac{dT}{dt} &+ \delta \frac{dv}{dt} - l_{v,21} \frac{d_i m^2}{dt} - l_{v,31} \frac{d_i m^3}{dt} \\
&= \frac{d}{dt}(q + w) - \sum_{k=0}^{3} \left(1 - \delta_1^k\right) \left(\frac{\partial e}{\partial m^k}\right)_{v,T} \frac{d_e m^k}{dt}
\end{aligned}
}
\tag{3.56}
$$

Since the coefficients pertaining to constant pressure (3.43) and constant volume (3.55) both have been derived from the first law, they will be related in some way. In order to establish this relationship, we express the specific volume in (3.56) as a function of the variables (p, T, m^k) so that dv can be developed according to (1.20) as

$$
\begin{aligned}
dv = \left(\frac{\partial v}{\partial p}\right)_{T,m^k} dp &+ \left(\frac{\partial v}{\partial T}\right)_{p,m^k} dT + (v_2 - v_1) d_i m^2 \\
&+ (v_3 - v_1) d_i m^3 + \sum_{k=0}^{3} (v_k - v_1) d_e m^k
\end{aligned}
\tag{3.57}
$$

Substitution of (3.57) into (3.56) gives

$$
\begin{aligned}
\left[c_v + \delta \left(\frac{\partial v}{\partial T}\right)_{p,m^k} \right] &\frac{dT}{dt} + \delta \left(\frac{\partial v}{\partial p}\right)_{T,m^k} \frac{dp}{dt} \\
- [l_{v,21} + \delta(v_1 - v_2)] \frac{d_i m^2}{dt} &- [l_{v,31} + \delta(v_1 - v_3)] \frac{d_i m^3}{dt} \\
= \frac{d}{dt}(q + w) - \sum_{k=0}^{3} &\left[\left(1 - \delta_1^k\right) \left(\frac{\partial e}{\partial m^k}\right)_{v,T} + \delta(v_k - v_1) \right] \frac{d_e m^k}{dt}
\end{aligned}
\tag{3.58}
$$

To find the desired relationships among the coefficients with reference to constant volume and constant pressure, we substitute the definitions of $(dw/dt, dq/dt, I^k, J^k)$ into (3.49) to obtain

$$
c_p \frac{dT}{dt} + \gamma \frac{dp}{dt} - l_{21} \frac{d_i m^2}{dt} - l_{31} \frac{d_i m^3}{dt} = \frac{d}{dt}(q + w) - \sum_{k=0}^{3} (h_k - h_1) \frac{d_e m^k}{dt}
\tag{3.59}
$$

Comparison of (3.59) and (3.58) then yields for $k \neq 1$

$$
\begin{aligned}
c_p &= c_v + \delta \left(\frac{\partial v}{\partial T}\right)_{p,m^k}, \qquad \gamma = \delta \left(\frac{\partial v}{\partial p}\right)_{T,m^k} \\
l_{21} &= l_{v,21} + \delta(v_1 - v_2), \qquad l_{31} = l_{v,31} + \delta(v_1 - v_3) \\
h_k - h_1 &= \left(\frac{\partial e}{\partial m^k}\right)_{v,T} + \delta(v_k - v_1)
\end{aligned}
\tag{3.60}
$$

The last expression also follows immediately by writing the internal energy in the form $e[v(p, T, m^k), T, m^k]$ so that the partial derivative may be expanded as

$$
\begin{aligned}
h_k - h_1 = e_k - e_1 + p(v_k - v_1) &= \left(\frac{\partial e}{\partial m^k}\right)_{p,T} + p(v_k - v_1) \\
&= \left(\frac{\partial e}{\partial m^k}\right)_{v,T} + \left(\frac{\partial e}{\partial v}\right)_{T,m^k}\left(\frac{\partial v}{\partial m^k}\right)_{p,T} + p(v_k - v_1) \\
&= \left(\frac{\partial e}{\partial m^k}\right)_{v,T} + \left(\frac{\partial e}{\partial v}\right)_{T,m^k}(v_k - v_1) + p(v_k - v_1) \\
&= \left(\frac{\partial e}{\partial m^k}\right)_{v,T} + \delta(v_k - v_1)
\end{aligned}
\tag{3.61}
$$

3.7 Summary

The first law of thermodynamics was introduced in the extensive form (3.4) stating that the change of the internal energy dE of the system is caused by heat transfer and work done on or by the system. This form of the first law is suitable for systems which are closed with respect to mass. By means of an example we have shown that the intensive form (3.8) is a suitable formulation for systems closed and open with respect to mass. The enthalpy h was introduced and the first law could be formulated in terms of this state function. Furthermore, it was shown that the partial derivatives of h with respect to T, p, and m^k resulted in the definitions of specific heat at constant pressure, specific heat of tension and the latent heat for phase transformations of the water substance. With the help of the Clausius–Kirchhoff relations additional relations between physical coefficients were derived. Finally, the heat equation was reformulated to give suitable forms of the prognostic equation for atmpheric temperature either involving p or v. Due to computer limitations the complex versions (3.52) and (3.56) have not yet been used in atmospheric modeling. Simplified versions of the heat equations will be discussed later.

3.8 Appendix A: dyadics in the cartesian system

Consider two vectors $\mathbf{A}$ and $\mathbf{B}$. The general or *dyadic product* is written as $\mathbf{AB}$ (no dot or cross between the two vectors). We define this product in terms of the algebra which is associated with it. The quantity $\mathbf{AB}$ is called a *dyad*, and the sum of two or more dyads is a *dyadic*. We postulate that for any vector $\mathbf{C}$ we have

$$
\mathbf{C} \cdot (\mathbf{AB}) = (\mathbf{C} \cdot \mathbf{A})\mathbf{B}, \qquad (\mathbf{AB}) \cdot \mathbf{C} = \mathbf{A}(\mathbf{B} \cdot \mathbf{C})
\tag{3.62}
$$

The dyad **AB** in expanded form can be written as follows:

$$
\begin{aligned}
\mathbf{AB} &= (\mathbf{i}A_x + \mathbf{j}A_y + \mathbf{k}A_z)(\mathbf{i}B_x + \mathbf{j}B_y + \mathbf{k}B_z) \\
&= \mathbf{ii}A_x B_x + \mathbf{ij}A_x B_y + \mathbf{ik}A_x B_z \\
&\quad + \mathbf{ji}A_y B_x + \mathbf{jj}A_y B_y + \mathbf{jk}A_y B_z \\
&\quad + \mathbf{ki}A_z B_x + \mathbf{kj}A_z B_y + \mathbf{kk}A_z B_z
\end{aligned}
\tag{3.63}
$$

If we denote the dyad **AB** by the symbol $\mathbb{D}$ we may write this dyad in the form

$$
\begin{aligned}
\mathbb{D} &= \mathbf{ii}D_{xx} + \mathbf{ij}D_{xy} + \mathbf{ik}D_{xz} \\
&\quad + \mathbf{ji}D_{yx} + \mathbf{jj}D_{yy} + \mathbf{jk}D_{yz} \\
&\quad + \mathbf{ki}D_{zx} + \mathbf{kj}D_{zy} + \mathbf{kk}D_{zz}
\end{aligned}
\tag{3.64}
$$

The dyad is called symmetric if $D_{xy} = D_{yx}$, $D_{xz} = D_{zx}$ and $D_{yz} = D_{zy}$.

The unit dyadic $\mathbb{E}$ is defined by

$$
\mathbb{E} = \mathbf{ii} + \mathbf{jj} + \mathbf{kk}
\tag{3.65}
$$

It plays the same prominent role in the theory of dyadics as the unit vector in vector analysis.

By forming the product of an arbitrary vector **A** or a dyad $\mathbb{D}$ with the unit dyadic we obtain

$$
\begin{aligned}
\mathbf{A} \cdot \mathbb{E} = \mathbf{A}, \qquad \mathbb{E} \cdot \mathbf{A} = \mathbf{A} \\
\mathbb{D} \cdot \mathbb{E} = \mathbb{D}, \qquad \mathbb{E} \cdot \mathbb{D} = \mathbb{D}
\end{aligned}
\tag{3.66}
$$

In general, the scalar product of a vector and a dyad is not commutative unless the dyad is symmetric. Hence we may write

$$
\begin{aligned}
\mathbf{A} \cdot \mathbb{D} \neq \mathbb{D} \cdot \mathbf{A}, \qquad \mathbb{D} \text{ is not symmetric} \\
\mathbf{A} \cdot \mathbb{D} = \mathbb{D} \cdot \mathbf{A}, \qquad \mathbb{D} \text{ is symmetric}
\end{aligned}
\tag{3.67}
$$

Sometimes it may be of advantage to use the definition of the double scalar product as expressed in

$$
\begin{aligned}
\mathbf{A} \cdot \mathbb{D} \cdot \mathbf{B} = \mathbf{A} \cdot (\mathbb{D} \cdot \mathbf{B}) = (\mathbb{D} \cdot \mathbf{B}) \cdot \mathbf{A} = \mathbb{D} \cdot\cdot \, \mathbf{BA} \\
\mathbf{A} \cdot \mathbb{D} \cdot \mathbf{B} = (\mathbf{A} \cdot \mathbb{D}) \cdot \mathbf{B} = \mathbf{B} \cdot (\mathbf{A} \cdot \mathbb{D}) = \mathbf{BA} \cdot\cdot \, \mathbb{D}
\end{aligned}
\tag{3.68}
$$

The operations are easy to understand by realizing that the scalar product of a vector and a dyad gives a vector. Moreover, the double scalar product is commutative.

3.9 Appendix B: Pfaff's differential form

The mathematical description of changes in physical systems often leads to expressions which are sometimes called *Pfaff's differential form*

$$
a_1 dx^1 + a_2 dx^2 + \cdots + a_m dx^m = \sum_{k=1}^{m} a_k dx^k = a_n dx^n
\tag{3.69}
$$

where use has been made of the repeated sum rule. The independent variables might be $x^k = p, T, m^0, m^2, m^3$ and the a_k are functions of the x^k. Whenever it is possible to set the differential expression (3.69) equal to the differential dy of a function $y = y(x^1, x^2, \ldots, x^m)$, (3.69) is said to be an *exact differential expression* and we can write

$$dy = \sum_{k=1}^{m} \left(\frac{\partial y}{\partial x^k}\right)_{x^j} dx^k = a_n dx^n \tag{3.70}$$

In the partial derivatives all x^j are held constant excepting the derivative with respect to x^k. Since the x^k are independent we find

$$a_k = \left(\frac{\partial y}{\partial x^k}\right)_{x^j}, \qquad k = 1, \ldots, m \tag{3.71}$$

The a_k and x^k are said to be conjugate to each other. It follows that if the differential expression (3.69) is exact, the a_k are partial differential coefficients. If y and its derivatives are continuous, then for any pair of the independent variables we have the mathematical requirement that

$$\frac{\partial^2 y}{\partial x^k \partial x^l} = \frac{\partial^2 y}{\partial x^l \partial x^k} \tag{3.72}$$

or

$$\left(\frac{\partial a_l}{\partial x^k}\right)_{x^j} = \left(\frac{\partial a_k}{\partial x^l}\right)_{x^j} \tag{3.73}$$

This equation holds for any two pairs of conjugate variables (a_i, x^i) and (a_j, x^j) in an exact differential expression and represents a necessary and sufficient condition for the exactness of (3.69).

3.10 Problems

3.1: Show the validity of the following expressions:

(a)
$$\left(\frac{\partial \psi}{\partial T}\right)_p = \left(\frac{\partial \psi}{\partial T}\right)_v + \left(\frac{\partial \psi}{\partial v}\right)_T \left(\frac{\partial v}{\partial T}\right)_p$$

(b)
$$\left(\frac{\partial \psi}{\partial p}\right)_T = \left(\frac{\partial \psi}{\partial v}\right)_T \left(\frac{\partial v}{\partial p}\right)_T$$

(c)
$$\left(\frac{\partial \psi}{\partial m^k}\right)_{p,T} = \psi_k - \psi_1 = \left(\frac{\partial \psi}{\partial m^k}\right)_{T,v} + \left(\frac{\partial \psi}{\partial v}\right)_T (v_k - v_1), \qquad k = 0, 2, 3$$

3.2: Verify the following identities:

$$\left(\frac{\partial c_v}{\partial v}\right)_{T,m^k} = \left(\frac{\partial(\delta - p)}{\partial T}\right)_v, \qquad \left(\frac{\partial c_v}{\partial m^k}\right)_{T,v} = -\frac{\partial l_{v,k1}}{\partial T}$$

3.3: For atmospheric systems we may assume $(\partial h/\partial p)_{T,m^k} = 0$ and $(\partial e/\partial v)_{T,m^k} = 0$. Show that $(\partial c_p/\partial p)_{T,m^k} = 0$ and $(\partial l_{k,1}/\partial p)_{T,m^k} = 0$.

3.4: Since $m^1 = 1 - (m^0 + m^2 + m^3)$ we can write instead of $e(T, p, m^0, m^1, m^2, m^3)$ the functional relationship $e(T, p, m^0, m^2, m^3)$. For fixed T and p find $d_i e$ for $e(T, p, m^0, m^1, m^2, m^3)$ and for $e(T, p, m^0, m^2, m^3)$. Show the difference, if any, between these two expressions.

3.5: Atmospheric pressure and internal energy per unit mass are given as functions of temperature and specific volume, that is $p = p(T, v)$ and $e = e(T, v)$.

(a) Show that by a change of coordinates we may express the temperature and the specific volume as functions of the internal energy and pressure, i.e. $T = T(e, p)$ and $v = v(e, p)$.
(b) Find the following partial derivatives

$$\left(\frac{\partial T}{\partial p}\right)_e, \quad \left(\frac{\partial v}{\partial p}\right)_e, \quad \left(\frac{\partial T}{\partial e}\right)_p, \quad \left(\frac{\partial v}{\partial e}\right)_p$$

to be expressed in terms of

$$\left(\frac{\partial p}{\partial T}\right)_v, \quad \left(\frac{\partial p}{\partial v}\right)_T, \quad \left(\frac{\partial e}{\partial T}\right)_v, \quad \left(\frac{\partial e}{\partial v}\right)_T$$

3.6: Experiments show that for gases $(\partial p/\partial T)_v \neq 0$. Using the fact that the internal energy is a function of state, show that the heat $đq$ added to the system is not a function of state.

3.7:

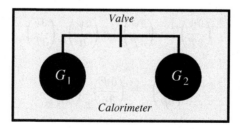

Fig. 3.4 Two metal containers G_1 and G_2 placed in a water calorimeter.

Two metal containers G_1, G_2 are placed in a water calorimeter as shown in Figure 3.4. G_1 contains a gas at high pressure, G_2 is perfectly evacuated. The temperature is uniform in the calorimeter. The valve between G_1 and G_2 is opened so that a pressure equalization takes place. It is observed that the temperature in the calorimeter does not change. You may assume that the internal energy is at most a function of the temperature and the specific volume.

What do you learn from this experiment?

3.8: Determine whether the following differential expressions are exact or not. If they are, find the functions of which these expressions are differentials. By inspection find integrating factors IF for the differentials which are not exact.

$$\text{(a)} \quad xdx + ydy, \qquad \text{(b)} \quad 2xydx - x^2dy,$$
$$\text{(c)} \quad xdx - ydy, \qquad \text{(d)} \quad ydx - xdy$$

3.9: Consider the function $p = p(T, v)$. Express p as a total differential and introduce the material coefficients

$$\beta = \frac{1}{p}\left(\frac{\partial p}{\partial T}\right)_v, \qquad \frac{1}{\kappa} = -v\left(\frac{\partial p}{\partial v}\right)_T.$$

Let β and κ be given by

$$\beta = \frac{1}{T} + \frac{a}{RvT^2}, \qquad \kappa = \frac{RTv(v-b)}{p(RTv^2 - av + ab)}$$

where a, b and R are constants. Find a thermal equation of state by integrating in the (v, T)-plane from the initial state (v_0, T_0) to the state (v, T).

Hint: The integration is independent of the path so that a particularly convenient integration path can be selected. Show the integration path in the (v, T)-plane.

3.10: An ideal gas is enclosed in a heat reservoir of temperature T_0. The heat exchange between the gas and the reservoir may be expressed by $đq/dt = a(T - T_0)$ where a is a constant. The volume of the flexible gas container enclosing the unit mass is given as a function of time, simply denoted by $v(t)$. At time $t = 0$ the volume and the temperature of the gas are denoted by v_i and T_i.

(a) Find an expression for the temperature of the gas as a function of time.
(b) Consider the special case $a = 0$.

3.11: The thermal equation of state is assumed to be given by $p = (RT/v)\exp[a/(vRT)]$ where a and R are constants. Determine the internal energy per unit mass e as a function of the temperature and the specific volume by performing a path integration in the (v, T)-plane from the initial state (v_0, T_0).

Stating it differently: find the function $f(T)$ in the expression $e(v, T) = e(v_0, T_0) + f(T)$. Since e is a state function you may choose a particularly convenient path.
   Given are:

$$\left(\frac{\partial e}{\partial T}\right)_v = c_v = \text{constant}, \qquad \left(\frac{\partial e}{\partial v}\right)_T = T\left(\frac{\partial p}{\partial T}\right)_v - p$$

4

The second law of thermodynamics

4.1 Introduction

It is well known that all natural physical processes are irreversible. Three brief examples will demonstrate this.

(i) It has never been observed that heat flowing from a warmer to a colder system will suddenly change its direction and flow from the colder to the warmer system. Nevertheless, the first law of thermodynamics does not prohibit this reversal of direction.

(ii) Consider a system consisting of two chambers. One of these is filled with a gas, the second chamber is completely evacuated. If the separating wall is pierced, a mass flow will take place until the pressure in both chambers is the same. It has never been observed that the original situation was restored by a return flow.

(iii) A stone is dropped into a water container resulting in an increase of the internal energy of the water container and, therefore, of its temperature. It never happens that the water container cools off spontaneously using the change of the internal energy to expel the stone.

In case of irreversible processes, the original state can be restored only by means of interactions with other systems which then suffer a remaining change. For example, to restore the original system in the second example, energy in form of work is required to evacuate the second chamber.

Irreversible processes taking place in isolated systems run in one direction only and thus provide the possibility of discerning between the past, the present and the future. In the past, the isolated system is farther removed from *thermodynamic equilibrium* than in the present and in the future the system is closer to thermodynamic equilibrium than at any previous time. Therefore, irreversible changes can be used to define a direction in time.

These examples imply that for the quantitative description of natural processes a suitable measure of the irreversibility is required. Such a measure is known as the entropy S.

4.2 Entropy

So far this measure of irreversibility has been given a name only. To be a useful physical quantity, it is necessary to provide a mathematical description. We assume that entropy such as internal energy is a scalar extensive quantity and can be balanced according to (2.5), i.e.

$$dS = d_iS + d_eS \tag{4.1}$$

meaning that within the interior of the system entropy is produced or destroyed (d_iS) or it flows into the system or out of it (d_eS). The decisive question is how the irreversibility can be expressed. We shall now briefly elaborate this question.

As in example (i) we consider two systems of differing temperatures. The two individual systems are connected by a conducting wall to form a new single system which will be completely isolated from its surroundings so that $d_eS = 0$. Heat will now flow within the system until the temperature equalization is complete and *thermal equilibrium* is achieved. Since the heat flow is an irreversible process the entropy of the system must change, i.e. $dS = d_iS \neq 0$. The fact that the irreversible process runs in one direction only is equivalent to the statement that the entropy must also change in one direction only. If in the isolated system the entropy change could be either positive or negative, then the entropy would not represent a unique measure for the deviation of the system from thermal equilibrium. It is conventional to assign an entropy increase for each *irreversible process* running in an isolated system, i.e.

$$d_iS \geq 0 \tag{4.2}$$

In case of a *reversible process* the equal sign applies meaning that the entropy of the isolated system remains constant.

The next step is to investigate the connection between entropy and energy. For this purpose we recall example (iii) where the temperature of the system has slightly increased due to the collision of the stone with the water and the bottom of the container. This change of state can be reversed only by an external action where the system is brought in thermal contact with an infinite heat reservoir having the original temperature of the system. Due to this external intervention the small amount of heat dQ will flow to the reservoir thus returning the system to its original thermodynamic state. After the temperature equalization all traces of the irreversible process have been removed and the entropy of the system resumes its initial value. The transfer of entropy from the system to the reservoir must, therefore, be proportional to $d\!\!\!{}^-Q$ and involve the temperature. The accepted mathematical statement of this physical process for the transfer of an infinitesimal quantity of heat is given by

$$d_e S = \frac{d\!\!\!{}^-Q}{T} \tag{4.3}$$

The amount of heat $d\!\!\!{}^-Q$ will be exchanged by a reversible process. In summary we can state that

$$dS = d_i S + d_e S$$

$$d_e S = \frac{d\!\!\!{}^-Q}{T}, \qquad d_i S \begin{cases} =0 & \text{reversible process} \\ >0 & \text{irreversible process} \end{cases} \tag{4.4a}$$

For isolated systems evidently we have

$$d_e S = 0 \implies dS = d_i S \geq 0 \tag{4.4b}$$

The equations collected in (4.4) are the mathematical statement of the second law of thermodynamics. As a next step we will restate the second law in a form most useful for atmospheric systems which can be open and inhomogeneous so that (4.3) must be generalized to include mass transfer also.

4.3 The axiomatic statements of the second law

We will now present three basic statements from which any other information concerning the second law will be deduced.

(i) The entropy S is an extensive variable of state.

All previously derived statements for Ψ and ψ also apply to S and s:

$$\boxed{S = \int_V \rho s\, d\tau, \qquad s = s_n m^n} \tag{4.5a}$$

From this we can form a budget equation by using the methods of Chapter 2. The integral form of the budget is then given by

$$\boxed{\frac{dS}{dt} = \int_V \rho \frac{ds}{dt} d\tau = -\int_\sigma \mathbf{F}_s \cdot d\mathbf{S} + \int_V Q_s d\tau} \tag{4.5b}$$

where Q_s is the entropy source and $\mathbf{F}_s$ is the nonconvective entropy flux. In general, $\mathbf{F}_s$ contains heat and mass transport. In an isolated system $\mathbf{F}_s \cdot d\mathbf{f} = 0$. The differential form of (4.5b) follows immediately from the divergence theorem:

$$\boxed{\rho \frac{ds}{dt} + \nabla \cdot \mathbf{F}_s = Q_s} \tag{4.5c}$$

(ii) The entropy production within the system is never negative.

$$
Q_s \begin{cases} = 0 & \text{reversible process} \\ > 0 & \text{irreversible process} \end{cases}
\tag{4.5d}
$$

(iii) The entropy change due to heat transport between the system and the surroundings at temperature T is given by

$$
\frac{d_{e,h}S}{dt} = -\oint_\sigma \frac{\mathbf{J}_s^h}{T} \cdot d\mathbf{S} = -\int_V \nabla \cdot \left(\frac{\mathbf{J}_s^h}{T}\right) d\tau
$$

$$
\text{or} \quad \rho \frac{d_{e,h}s}{dt} = -\nabla \cdot \left(\frac{\mathbf{J}_s^h}{T}\right)
\tag{4.5e}
$$

These three axiomatic statements form the second law of thermodynamics. The sensible heat flux $\mathbf{J}_s^h$ divided by T is known as the *reduced sensible heat flux*. Since the entropy is an extensive variable of state, in open systems entropy changes are induced not only by heat transfer but also by diffusive mass fluxes. We denote the external entropy change resulting from the diffusive mass fluxes by $d_{e,d}s$. Then in an open system the total change of entropy due to interactions with the surroundings may be written as

$$
\frac{d_e s}{dt} = \frac{d_{e,h}s}{dt} + \frac{d_{e,d}s}{dt} = -v\nabla \cdot \left(\frac{\mathbf{J}_s^h}{T}\right) - v\nabla \cdot (\mathbf{J}^n s_n) = -v\nabla \cdot \left(\frac{\mathbf{J}_s^h}{T} + \mathbf{J}^n s_n\right)
\tag{4.6}
$$

where the diffusive entropy flux $\mathbf{J}^n s_n$ has been introduced in analogy to the diffusive enthalpy flux $\mathbf{J}^n h_n$.

In the following section we are going to verify the second part of this equation with the help of Gibbs' fundamental equation. This important equation is the starting point for the derivation of the entropy production equation from which numerous decisive conclusions follow. Moreover, the reduced sensible heat flux will also appear as a part of Gibbs' fundamental equation.

4.4 Gibbs' fundamental equation and entropy production

To make things simple, let us first consider a one-component system. The Gibbs' equation, in fact, is a combination of the first and the second law of thermodynamics. We proceed with the derivation of this equation by replacing the right-hand side of the first law with the help of (4.4a), but we include $d\hspace{-0.3em}^{\bar{}}w$ in the definition of the entropy change. Thus we find

$$de + pdv = đq + đw = Tds$$ (4.7)

which, by means of a Legendre transformation, can also be rewritten in the form

$$d(e + pv - Ts) - vdp + sdT = 0$$ (4.8)

The expression in parentheses is known as the specific Gibbs function μ

$$\mu = e + pv - Ts = h - Ts$$ (4.9)

Substitution of (4.9) into (4.8) results in

$$d\mu = vdp - sdT$$ (4.10)

from which we recognize that the Gibbs function depends on pressure p and temperature T, i.e. $\mu = \mu(p, T)$. The fact that the variables (p, T) are reported in routine observations, makes this function so valuable for atmospheric thermodynamics.

The differential statement of the Gibbs function

$$d\mu = \left(\frac{\partial\mu}{\partial p}\right)_T dp + \left(\frac{\partial\mu}{\partial T}\right)_p dT$$ (4.11)

and the comparison with (4.10) results in the important identities

$$\left(\frac{\partial\mu}{\partial p}\right)_T = v, \qquad \left(\frac{\partial\mu}{\partial T}\right)_p = -s$$ (4.12)

We will now generalize (4.11) by considering a multi-component system so that $\mu = \mu(p, T, m^k)$. According to (1.18) the differential $d\mu$ can now be written as

$$d\mu = \left(\frac{\partial\mu}{\partial p}\right)_{T,m^k} dp + \left(\frac{\partial\mu}{\partial T}\right)_{p,m^k} dT + \mu_n dm^n = vdp - sdT + \mu_n dm^n$$ (4.13)

with

$$\mu_k = h_k - s_k T$$ (4.14)

The quantity μ_k is called the *chemical potential*. The identities (4.12) are still valid, but now the concentrations m^k must be held fixed in addition to T or p. We will return to these identities in Chapter 6. A suitable Legendre transformation of (4.13) gives

$$d(\mu + sT) - Tds = vdp + \mu_n dm^n$$ (4.15)

from which we obtain two equivalent forms of Gibbs' fundamental equation

$$T ds = dh - v dp - \mu_n dm^n = de + p dv - \mu_n dm^n = d(q + w) - \mu_n dm^n$$
(4.16)

As a prognostic equation this relation can also be written as

$$\boxed{T \frac{ds}{dt} = \frac{d(q + w)}{dt} - \mu_n \frac{dm^n}{dt}}$$
(4.17)

which will be used to find a suitable expression for the entropy production.

Gibbs' fundamental equation was derived for the case of equilibrium thermo-dynamics where the state of a given system is assumed to be completely specified by a finite number of state variables which is usually quite small for atmospheric problems. Analogously, in case that irreversible processes are taking place, we also assume that the state of the system is completely specified by a finite number of fields of state variables that are functions of space and time. For atmospheric problems the number of fields is usually quite small. A basic assumption in the work to follow is that Gibbs' equation may also be applied to problems of nonequilibrium, but the system must not be too far removed from equilibrium.

Let us now subdivide the system into tiny mass elements which must still be large enough to contain a large number of atoms. In this case each mass element may be treated as a thermodynamic system. Moreover, at any time each individual mass element should be in thermodynamic equilibrium so that local equilibrium exists everywhere. If this condition is satisfied, for all fields of state variables the numerous relations of equilibrium thermodynamics are still valid. However, the analysis of irreversible thermodynamics is no longer valid when physical processes proceed so rapidly that large inhomogeneities form in the fluid, since in this case thermodynamic state variables can no longer be defined. With the help of the kinetic theory of transport processes it is possible to estimate the limits of validity of local thermodynamic equilibrium. In the case that temperature plays a dominating role, local thermodynamic equilibrium may be assumed as long as temperature changes over distances defined by the mean free path are small in comparison to the temperature itself. For air at standard atmospheric conditions the mean free path is of the order of magnitude of 10^{-6} to 10^{-7} m.

In the following we formulate the entropy production equation which will be used later to obtain suitable expressions for the fluxes appearing in the first law of thermodynamics as stated, for example, in (3.52). One of these fluxes is the radiative flux F_R which will be omitted from further considerations involving the entropy production. The reason for this is that this flux is best evaluated with the help of the theory of radiative transfer. Moreover, it is very difficult and inconvenient to find F_R with the help of the irreversible theory. Callies and Herbert (1984), Herbert and

Pelkowski (1990), however, have shown how to proceed in case of heat radiation by investigating the entropy of the radiation field.

We are now ready to continue the mathematical development. For $đq/dt$ we substitute

$$\frac{đq}{dt} = -v\nabla \cdot \mathbf{J}^h \tag{4.18}$$

and for $đw/dt$

$$\frac{đw}{dt} = v\mathbb{J}\cdot\cdot\nabla\mathbf{v} \tag{4.19}$$

The expression dm^k/dt is replaced with the help of the budget equation (2.17). It will be left to the exercises that after a few easy steps we find

$$\rho\frac{ds}{dt} = -\nabla \cdot \left(\frac{\mathbf{J}_s^h}{T} + s_n\mathbf{J}^n\right) - \frac{\mu_n}{T}I^n - \frac{\mathbf{J}^h}{T^2}\cdot\nabla T - \mathbf{J}^n \cdot \nabla\left(\frac{\mu_n}{T}\right) + \frac{1}{T}\mathbb{J}\cdot\cdot\nabla\mathbf{v} \tag{4.20}$$

which contains the reduced sensible heat flux $\mathbf{J}_s^h$ and the entropy transport term $s_n\mathbf{J}^n$.

From (4.5b) we conclude that the entropy change d_eS due to the system's inter-action with the surroundings can be expressed with the help of a surface integral over the flux or by the divergence of this flux. The entropy change d_iS within the system due to irreversible processes is never negative and cannot be expressed by a surface integral over a flux or by the divergence of this flux. We accept Gibbs' fundamental equation in the useful form (4.20) and carry out a comparison with the general form of the budget equation for the entropy (4.5c). Thus we find that the nonconvective flux $\mathbf{F}_s$ is given by

$$\mathbf{F}_s = \frac{\mathbf{J}_s^h}{T} + s_n\mathbf{J}^n \tag{4.21}$$

and the source Q_s by

$$TQ_s = -\mu_n I^n - \frac{\mathbf{J}^h}{T}\cdot\nabla T - T\mathbf{J}^n \cdot \nabla\left(\frac{\mu_n}{T}\right) + \mathbb{J}\cdot\cdot\nabla\mathbf{v} \geq 0 \tag{4.22}$$

where, according to (4.5d), $TQ_S > 0$ denotes irreversible processes.

For our purposes it will be expedient to express the entropy production equation (4.22) in a still more suitable form by introducing the *chemical affinities* a_{k1}. They are defined as the differences of the chemical potentials

$$\begin{aligned}\left(\frac{\partial\mu}{\partial m^2}\right)_{p,T} &= \mu_2 - \mu_1 = a_{21} \\ \left(\frac{\partial\mu}{\partial m^3}\right)_{p,T} &= \mu_3 - \mu_1 = a_{31} \\ \mu_3 - \mu_2 &= a_{32}\end{aligned} \tag{4.23}$$

In order to introduce the a_{k1} we need to reformulate the expression $\mu_n I^n$ in (4.22) by replacing the phase transition flux I^1. Since $I^0 = 0$ we find $I^1 = -(I^2 + I^3)$ so that (4.22) may be expressed as

$$TQ_s = -a_{21}I^2 - a_{31}I^3 + T\mathbf{J}^h \cdot \nabla\left(\frac{1}{T}\right) - T\mathbf{J}^n \cdot \nabla\left(\frac{\mu_n}{T}\right) + \mathbb{J} \cdot\cdot \nabla \mathbf{v} \geq 0$$

$$(4.24)$$

This is the form of the entropy production equation which will be used to obtain the proper form of fluxes appearing in the first law of thermodynamics.

Each term in the entropy production equation is the product of two factors. One factor represents the *irreversible flux* (I^k, $\mathbf{J}^k$, $\mathbf{J}^h$, $\mathbb{J}$) while the other factor is the so-called *conjugated thermodynamic force* driving the flux. The specification of the irreversible fluxes will be treated in great detail in a later chapter. At this point it will be sufficient to make a few general observations.

(i) From (4.24) it is seen that entropy is produced by phase transition rates, by dissipative processes and by processes resulting from nonvanishing gradients of the thermodynamic variables.

(ii) Reference to (3.31) shows that the term dk_d/dt was suppressed. Had this term been carried along, then Q_s would contain an additional term of the form $\mathbf{J}^n \cdot d\mathbf{v}_{n,d}/dt$. This means that such irreversible processes would be admitted which tend to equalize the diffusion velocity.

We consider now the artificial situation that phase changes are the only irreversible processes taking place. This can be realized only if the system is homogeneous with respect to the thermodynamic variables and if viscosity can be totally ignored. In this case the entropy production equation reduces to

$$vTQ_s = -a_{21}\frac{d_i m^2}{dt} - a_{31}\frac{d_i m^3}{dt} \geq 0 \qquad (4.25)$$

In order to get a clearer understanding about the physical meaning of the affinities we will consider the example of a water cloud which is characterized by $m^3 = 0$ and $d_i m^3/dt = 0$ so that (4.25) is further reduced to

$$vTQ_s = -a_{21}\frac{d_i m^2}{dt} \geq 0 \qquad (4.26)$$

Obviously, the sign of $d_i m^2/dt$ will be determined by the sign of a_{21}. If $a_{21} > 0$ then $d_i m^2/dt < 0$. This means that liquid water evaporates. In case that $a_{21} < 0$ then $d_i m^2/dt > 0$ so that water vapor condenses. Analogous results are obtained for an ice cloud with $m^2 = 0$ and $d_i m^2/dt = 0$.

As previously stated, all reversible processes are idealizations and at best can be approximated by quasi-static processes. In the atmosphere entropy is produced continually by various physical processes so that stationarity over extended periods of time can be sustained only if entropy is exported to the exterior world. We shall take up this topic in a later chapter.

4.5 Calculation of entropy changes

4.5.1 Entropy and reversibility

Let us consider the simplified process that the only interaction of the system with its surroundings is the reversible heat transfer due to a finite temperature difference $T_f - T_i$ where the subscripts f and i stand for final and initial. We may approximate such a process by the heat transfer between the system and a very large set of heat reservoirs ranging from T_i to T_f. A *heat reservoir* is a body of such a large mass that it may absorb or reject an unlimited quantity of heat without suffering a measurable change in temperature or any other thermodynamic coordinate. During any infinitesimal portion of the process the amount of heat transferred reversibly at temperature T between the system and one of the reservoirs is dQ_R which we consider a positive number. The subscript R stands for reversibility. In a simplified example the entropy change due to the heat transfer from the reservoir to the system is given by

$$\text{change of the system}: \quad dS_s = \frac{dQ_R}{T}$$
$$\text{change of the reservoir}: \quad dS_r = -\frac{dQ_R}{T} \tag{4.27}$$

The sum of the changes, loosely called the *entropy change of the universe*, amounts to zero. If the system rejects heat and the reservoir absorbs it, then dS_s and dS_r change sign so that the entropy change of the universe is zero again. Since this statement is true for a particular portion of the reversible process, it will be true for all other portions within the given temperature range so that the entropy change for the complete reversible process is zero also. This is the same as saying that the entropy of the "universe" remains constant.

4.5.2 Entropy and irreversibility

We consider a system undergoing an irreversible process between an initial equilibrium state and a final equilibrium state. In order to calculate the entropy change, which by necessity increases the entropy, ($\Delta S_I > 0$), we make use of the fact that the entropy is a variable of state so that the entropy change does not depend on the path. Therefore, we are permitted to replace the original irreversible path I by a suitable reversible path R which takes the system from the given initial state, i to the final state, f of the irreversible path. Thus, the entropy change ΔS_I of the irreversible process is now replaced by

$$\Delta S_R = \int_{R\,i}^{f} \frac{dQ}{T} \tag{4.28}$$

with $\Delta S_I = \Delta S_R$. When either the initial or the final state is a nonequilibrium state, we can still calculate the entropy difference by special methods. We will demonstrate the two situations by two relatively simple examples.

4.5.2.1 Entropy change between two equilibrium states

Let us consider the very important *Gay-Lussac experiment* showing that the internal energy of an ideal gas depends on temperature only. We will split this example into three parts. In the first part (i) we will derive the ideal gas law. In the second part (ii) we will show that the internal energy of an ideal gas depends on temperature only. In the last part (iii) we will calculate the entropy change of the Gay-Lussac experiment.

(i) Every ideal gas obeys two basic equations which are due to *Boyle* and to *Gay-Lussac*:

$$\text{Boyle :} \quad pV = \text{constant} \quad \text{for constant temperature} \tag{4.29}$$

$$\text{Gay-Lussac:} \quad V/T = \text{constant} \quad \text{for constant pressure} \tag{4.30}$$

These two laws will now be combined to give the ideal gas law. From (4.29) and (4.30) it follows immediately that the general functional form of the equation of state of an ideal gas can be written as

$$F(p, V, T) = 0 \tag{4.31}$$

Since in (4.29) the right-hand side depends on T only, in (4.30) only on p, we may write

$$pV = f_1(T) \quad \text{and} \tag{4.32}$$

$$\frac{V}{T} = f_2(p) \tag{4.33}$$

We now eliminate V and obtain

$$pf_2(p) = \frac{f_1(T)}{T} \tag{4.34}$$

The left-hand side depends only on p, the right-hand side only on T. Therefore, both sides must equal a constant K

$$pf_2(p) = \frac{f_1(T)}{T} = K \tag{4.35}$$

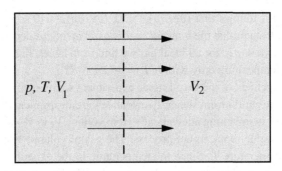

Fig. 4.1 Irreversible expansion, Gay-Lussac experiment.

Substituting $f_1(T)$ from (4.35) into (4.32) gives

$$pV = KT \tag{4.36}$$

According to *Avogadro's law*, all ideal gases assume the same molar volume for the same pressure and temperature. Replacing the volume V in (4.36) by the molar volume $\breve{v}$ results in the *universal gas law for ideal gases*,

$$p\breve{v} = R^*T \tag{4.37}$$

where K has been replaced by the *universal gas constant* R^* which, by definition, refers to one mole. If V is the volume of N moles ($V = \breve{v}N$), we may also write

$$pV = NR^*T \tag{4.38}$$

For meteorological purposes it is advantageous to introduce the specific volume v. Denoting the molecular weight by $\breve{m} = M/N$ and the *individual gas constant* by $R = R^*/\breve{m}$, the ideal gas law is now written as

$$pv = RT \tag{4.39}$$

(ii) An adiabatically isolated chamber ($đQ = 0$) consists of two compartments of volumes V_1 and V_2, see Figure 4.1. Compartment 1 is filled with an ideal gas held at pressure p and temperature T. Compartment 2 is completely evacuated, so that no external work is done by the gas when the partition separating the compartments is broken, $đA_{\rm E} = 0$. Measurements show that the temperature before and after the expansion is the same.

According to the first law, $dE = 0$. For an ideal gas, reverting to the intensive notation, we may either write

$$e = e(p, T) \quad \text{or} \quad e = e(v, T) \implies$$
$$de = \left(\frac{\partial e}{\partial p}\right)_T dp + \left(\frac{\partial e}{\partial T}\right)_p dT = \left(\frac{\partial e}{\partial v}\right)_T dv + \left(\frac{\partial e}{\partial T}\right)_v dT = 0 \tag{4.40}$$

Since $dT = 0$ it follows that $(\partial e / \partial p)_T = 0$, $(\partial e / \partial v)_T = 0$ so that $e = e(T)$. This important result states that the *internal energy of an ideal gas* depends on temperature only. Since $h = e + pv$, for an ideal gas we find from (4.39) that the enthalpy is also a function of temperature only with $h(T) = e(T) + RT$.

(iii) From the arrangement of the Gay-Lussac experiment we recognize that the initial and the final state are equilibrium states. To calculate the entropy change of the irreversible process (the gas is streaming adiabatically from volume V_1 to V_2) we need to construct a suitable reversible replacement process. The gas of volume V_1 is placed in a heat reservoir of temperature T. Since the temperature in the experiment did not change, we expand the gas isothermally from V_1 to $V_1 + V_2$. The free expansion entropy change of the system can then be calculated very easily

$$\Delta S_R = \int_{R}^{f} \frac{dQ}{T} = \frac{1}{T} \int_{R}^{f} (dE + p\,dV) = NR^* \int_{R}^{V_1+V_2} \frac{dV}{V}$$
$$= NR^* \ln \frac{V_1 + V_2}{V_1} > 0 \tag{4.41}$$

4.5.2.2 Entropy change from a nonequilibrium to an equilibrium state

A thermally conducting bar of uniform cross-section A is placed between a hot and a cold reservoir. After a certain time, when the temperature distribution along the bar is nonuniform, the bar will be disconnected from the reservoirs, then thermally insulated and kept at constant pressure. An internal flow of heat will bring the bar to a uniform temperature distribution, but the transition will be from an initial nonequilibrium state to a final equilibrium state, see Figure 4.2.

It is impossible to construct a single reversible process by which the system – the bar – may be brought from the given initial to the final state. To calculate the entropy change, we think of the bar to be composed of an inifinite number

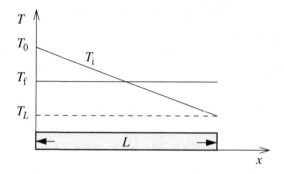

Fig. 4.2 Temperature distribution along the bar.

of infinitesimally thin slices. Each of these has a different initial temperature but all have the same final temperature. We imagine that all slices are insulated from one another and that each slice is brought in thermal contact successively with a series of reservoirs ranging from the initial to the final temperature. Suppose for simplicity that the initial nonuniform temperature distribution is linear in distance along the bar of length L. If the material properties of the bar are constants and no heat is lost to the surroundings, then the final temperature distribution will be T_f as shown in Figure 4.2. Since the initial temperature distribution T_i is linear we have $T_i = T_0 - x(T_0 - T_L)/L$.

Each single slice of thickness dx has the mass $dM = \rho A dx$ and the heat capacity $c_p dM$ so that $dQ = c_p\, dM\, dT$. The entropy change of each slice is then calculated directly from the definition of entropy (or what amounts to the same thing from Gibbs' equation in enthalpy form) and is given by

$$\Delta S = \int_i^f \frac{dQ}{T} = c_p\rho A\, dx \int_{T_i(x)}^{T_f} \frac{dT}{T} = -(c_p\rho A\, dx) \ln\left(\frac{T_0}{T_f} - \frac{T_0 - T_L}{LT_f}x\right)$$

(4.42)

The entire entropy change is then found by integrating over the length of the bar

$$\begin{aligned}
S_f - S_i &= -c_p\rho A \int_0^L \ln\left(\frac{T_0}{T_f} - \frac{T_0 - T_L}{LT_f}x\right) dx \\
&= C_p\left(1 + \ln T_f + \frac{T_L \ln T_L - T_0 \ln T_0}{T_0 - T_L}\right)
\end{aligned}$$

(4.43)

with $C_p = c_p\rho AL$. Substituting some suitable numbers for T_0 and T_L (say, $T_0 = 400$ K, $T_L = 200$ K, hence $T_f = 300$ K) results in $S_f - S_i = 0.019\, C_p > 0$ which is a positive quantity as expected.

4.6 Summary and additional remarks

We have discussed the entropy concept and found that $d_i S > 0$ whereas $d_e S \gtreqless 0$. The second law of thermodynamics was described by means of three axiomatic statements. An expression for the nonconvective entropy flux $\mathbf{F}_s$ was obtained. Gibbs' fundamental equation was introduced, the chemical potential and the chemical affinities were defined. The very important entropy production equation was derived for later use. Examples were given to demonstrate the calculation of entropy changes.

We have used the basic equations of Boyle and Gay-Lussac to derive the ideal gas law which can also be obtained with the help of a particle model. The model is discussed in detail in textbooks on statistical thermodynamics. The basic kinetic theory of an ideal gas includes the assumptions that the molecules exert no forces on each other except when they collide. Collisions between the molecules and the walls enclosing the gas are perfectly elastic so that there is no decrease of kinetic energy when the molecules collide. There is a more refined equation of state which is due to *van der Waals*. The theory leading to this equation admits the existence of forces between molecules. Moreover, the theory accounts for the actual volume of a molecule whereas the ideal gas molecules are point masses. Without further details we will state the van der Waals equation

$$(\breve{v} - b)\left(p + \frac{a}{\breve{v}^2}\right) = R^*T \tag{4.44}$$

Utilizing in this equation the specific volume v instead of the molar volume $\breve{v}$, the universal gas constant R^* must be replaced by the individual gas constant $R = R^*/\breve{m}$. The constant b accounts for the volume of the molecules while the quantity $a/\breve{v}^2$ denotes attractive forces between molecules.

The ideal gas law is sufficient for most meteorological purposes since the atmosphere may be considered a low-density gas with sufficiently high temperatures. The two gas laws we have stated are by no means the only ones. There exist more than 100 other equations of state which are of some importance to engineering but not to meteorology.

There exists also a *third law of thermodynamics* which is also known as *Nernst's theorem of heat*. In contrast to the zero-th, the first and the second law of thermodynamics, no new state variables are introduced. The third law of thermodynamics describes the behavior of entropy S_0 near the absolute zero temperature. Following Planck, this entropy value is $S_0 = 0$.

Clausius pointed out that the entropy increase in an isolated nonequilibrium system finally results in thermal equilibrium. If temperature differences cease to exist, then there can be no transformation of internal energy into mechanical work. Since an isolated system cannot be truly realized in the real world, Clausius extended the isolated system of finite size to the entire universe. He then deduced from (4.5) that the entropy of the universe approaches a maximum value. The final point of the development of the universe would be the so-called *Heat Death of the Universe*. However, new cosmological theories point out that the universe cannot be treated as an isolated thermodynamic system for which (4.5) holds and the hypothesis of the heat death of the universe is no longer acceptable.

We will conclude this little excursion with reference to a newer theory. While the entropy production for stationary states near equilibrium reaches a minimum value, entirely different situations exist for stationary states in open systems which happen to be far from equilibrium. In this nonlinear region it is possible that $dS < 0$ if the entropy export ($d_e S < 0$) exceeds the internal production ($d_i S > 0$). If such conditions exist, local microscopic fluctuations may occur and spread over the entire system causing a transition to a stationary state of higher structural order. This process, called *self-organization*, is of great importance for the interpretation of biological evolution.

As an example of self-organization, meteorology provides the convective instability which plays a fundamental role in convective cloud formation. From a fairly homogeneous temperature and moisture field over the ocean surface a well-organized cloud field with hexagonal cloud structure often develops. This type of structure formation was first observed in the laboratory and is known as the *Benard problem*.

4.7 Problems

4.1: One mole of oxygen undergoes a free expansion from a volume of 5 liters to a volume of 10 liters. The temperature of the gas is 293 K.

(a) What is the entropy change of the gas and of the universe?
(b) If the expansion is performed reversibly and isothermally, what is the entropy change of the oxygen and of the universe?

4.2: Ten grams of water at a temperature of 293 K are converted to ice at 263 K at constant atmospheric pressure. Assume that c_p of ice is 2.09 Joule g^{-1} K^{-1}. The heat of fusion of ice at 273 K is 334 Joule g^{-1}. What is the total entropy change of the system?

4.3: A mass M of water at T_1 is isobarically and adiabatically mixed with an equal mass of water at T_2. What is the entropy change of the universe? Show that this change is positive.

4.4: Consider an ideal gas. Assuming that c_v can be represented by a Taylor expansion, show that $\oint dq/T = 0$.

4.5: Show that for an isobaric-adiabatic process the enthalpy H assumes a minimum value.

4.6: Find expressions for the internal energy and the entropy of a van der Waals gas. The equation of state for this type of gas is given by (4.44). Hint: Show first that c_v depends on temperature only.

4.7: Assume the existence of the exact differential $df = -pdv - sdT$.

(a) Show that for a one-component system Gibbs' equation can be written in the following forms:

$$(1) \quad Tds = c_v dT + T\left(\frac{\partial p}{\partial T}\right)_v dv$$

$$(2) \quad Tds = c_p dT - T\left(\frac{\partial v}{\partial T}\right)_p dp$$

(b) Use equation (1) to determine the amount of heat which is transferred if one mole of a van der Waals gas undergoes a reversible isothermal expansion. The equation of state is given by (4.44).

4.8: Derive the identity

$$\left(\frac{\partial e}{\partial v}\right)_T = T\left(\frac{\partial p}{\partial T}\right)_v - p$$

and find $(\partial e/\partial v)_T$ for an ideal gas and for a van der Waals gas. The equation of state for the van der Waals gas is given by (4.44).

4.9: (a) Use the two Tds-equations of problem (4.7) to derive the following expression:

$$c_p - c_v = -T\left(\frac{\partial v}{\partial T}\right)_p^2 \left(\frac{\partial p}{\partial v}\right)_T$$

(b) Find $c_p - c_v$ for an ideal gas.

4.10: The entropy of saturated water at 100°C is listed in the entropy tables as 1.297 Joule g^{-1} K^{-1} and that of saturated steam for the same temperature as 7.366 Joule g^{-1} K^{-1}.

(a) What is the heat of vaporization at this temperature?
(b) The enthalpy of saturated steam at 100°C is listed in the enthalpy tables as 2.678 Joule g^{-1}. Find the enthalpy of saturated water at this temperature.
(c) Find the Gibbs function of saturated water and of saturated steam at this temperature. Show that the Gibbs functions are equal.

4.11: (a) The Massieu function or potential is defined by $J = -E/T + S$. Show that $dJ = (E/T^2)\,dT + (p/T)V$.
(b) The Planck function K is defined by $K = -H/T + S$. Show that $dK = H/T^2 dT - (V/T)dp$.

4.12: Obtain the potential functions $J(1/T, V) = -E/T + S$, $\Psi(E, p/T) = -pV/T + S$ and $K(1/T, p/T) = -G/T$ by means of Legendre transformations of Gibbs' fundamental equation. Show that the functional notation on the left-hand side of these equations is written correctly.

5

Thermal radiation

5.1 Introduction

In this chapter we will show how the laws of thermodynamics can be applied to obtain the important Stefan–Boltzmann radiation law and the entropy of the blackbody radiation. Before we proceed with thermodynamic reasoning, we need to define a very basic quantity from radiative transfer theory which is known as the *radiance*. The radiance L is the electromagnetic energy of the radiation field leaving or approaching a differential area $d\tilde{f}$ of an imaginary plane in the time interval dt. The direction of the energy propagation is defined by a differential solid angle $d\omega$ whose central direction Ω is normal to the imaginary plane, see Figure 5.1. Therefore, the dimension of the radiance is energy/(area $*$ time $*$ steradian).

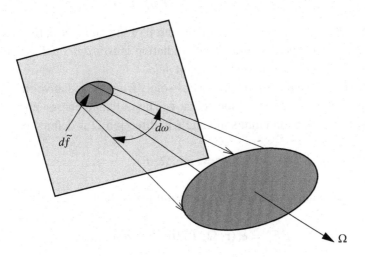

Fig. 5.1 Propagation of radiative energy.

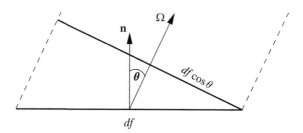

Fig. 5.2 Definition of radiance, $d\mathbf{f} = df\,\mathbf{n}$, $d\tilde{f} = df\cos\theta$.

We now consider the energy transport across an area element whose direction is defined by the unit normal $\mathbf{n}$. Since the direction of the unit vector Ω and $\mathbf{n}$ generally differ, the radiance must be referenced with respect to $\cos\theta\,df$ which is normal to Ω, as follows from Figure 5.2.

Therefore, the *radiative energy* is given by

$$dE = L(\mathbf{r}, \Omega)\,d\mathbf{f} \cdot \Omega\,d\omega\,dt = L(\mathbf{r}, \Omega)\cos\theta\,df\,d\omega\,dt \qquad (5.1)$$

In general L not only depends on the position vector $\mathbf{r}$ of the differential area and the direction of the radiation but also on the frequency. Let the radiance contained within the frequency interval between ν and $\nu + d\nu$ be defined by $L_\nu(\mathbf{r}, \Omega)$, then

$$L(\mathbf{r}, \Omega) = \int_0^\infty L_\nu(\mathbf{r}, \Omega)\,d\nu \qquad (5.2)$$

and

$$dE_\nu = L_\nu(\mathbf{r}, \Omega)\cos\theta\,df\,d\omega\,dt\,d\nu \qquad (5.3)$$

where explicit reference is made to the direction of the radiance at the position $\mathbf{r}$. If L is independent of the direction Ω, the radiation is *isotropic*, if L is independent of $\mathbf{r}$ we speak of a *homogeneous radiation field*.

We now consider a radiation field arising solely from the radiative energy emitted by a body K of temperature T located at $\mathbf{r}$. The radiative energy emitted by unit surface area, unit time and unit solid angle in direction Ω is defined by

$$e_\nu = e_\nu(\mathbf{r}, \Omega, T) \qquad (5.4)$$

This term is also called the *spectral emission* of the body. According to (5.3) we obtain

$$dE_\nu^{\mathrm{e}} = e_\nu(\mathbf{r}, \Omega, T)\cos\theta\,df\,d\omega\,dt\,d\nu \qquad (5.5)$$

so that e_ν is the radiance in the immediate neighborhood of df.

In addition to the spectral emission e_ν, a radiating surface may also absorb radiation. Let

$$dE_\nu^i = L_\nu(\mathbf{r}, \Omega) \cos\theta \, df \, d\omega \, dt \, d\nu \qquad (5.6)$$

represent the radiative energy incident on df. A part of this energy will be absorbed by the surface element, the remaining part is reflected. Let dE_ν^a represent that part of dE_ν^i which is absorbed so that

$$dE_\nu^a = a_\nu(\mathbf{r}, \Omega, T) \, dE_\nu^i \qquad (5.7)$$

The quantity a_ν is called the *spectral absorptivity* of the body. The reflected part of the incident radiative energy is then given by

$$
\begin{aligned}
dE_\nu^r = dE_\nu^i - dE_\nu^a &= dE_\nu^i \, [1 - a_\nu(\mathbf{r}, \Omega, T)] \\
&= [1 - a_\nu(\mathbf{r}, \Omega, T)] \, L_\nu(\mathbf{r}, \Omega) \, \cos\theta \, df \, d\omega \, dt \, d\nu
\end{aligned} \qquad (5.8)
$$

There exists a remarkable relationship between e_ν and a_ν in case of thermal radiation. This relationship is known as Kirchhoff's law which is of great importance in radiative transfer theory.

5.2 Kirchhoff's law

Consider two arbitrary radiating bodies K_1 and K_2 of temperatures T_1 and T_2 which are surrounded by an adiabatic enclosure. If $T_1 > T_2$, then the amount of radiative energy which K_1 radiates to K_2 exceeds the amount of energy which K_1 receives from K_2. After a certain time K_1 and K_2 will have the same temperature so that *radiative equilibrium* is observed within the enclosure. Then K_1 emits as much energy as it receives from K_2, otherwise the radiative equilibrium would be disturbed. Let the radiative energy arriving at K_1 from K_2 be given by (5.7). If $a_{\nu,1}$ is the absorptivity of K_1, then it absorbs the energy

$$dE_\nu^a = a_{\nu,1} L_\nu \cos\theta \, df \, d\omega \, dt \, d\nu \qquad (5.9)$$

In case of radiative equilibrium, the absorbed and the emitted energy must be equal

$$dE_\nu^a = dE_\nu^e = e_{\nu,1} \cos\theta \, df \, d\omega \, dt \, d\nu \qquad (5.10)$$

so that

$$e_{\nu,1} = a_{\nu,1} L_\nu \qquad (5.11)$$

By an analogous argument involving K_2, we would have found

$$e_{\nu,2} = a_{\nu,2} L_\nu \qquad (5.12)$$

from which follows

$$\frac{e_{v,1}}{a_{v,1}} = \frac{e_{v,2}}{a_{v,2}} = L_v \tag{5.13}$$

If n bodies are enclosed, the same type of argument could be carried out, resulting in

$$\frac{e_{v,1}}{a_{v,1}} = \frac{e_{v,2}}{a_{v,2}} = \cdots = \frac{e_{v,n}}{a_{v,n}} = L_v \tag{5.14}$$

No assumptions have been made about the form and material composition of the radiating bodies so that their surface properties may vary from one body to the next. Therefore, e and a will also depend on the material properties of the radiators. It is important to realize that in case of radiative equilibrium the radiance L is of the same magnitude everywhere in the cavity and, therefore, must be independent of the material properties of the radiators. Hence, the material properties contained in e and a must cancel. This must be true for any frequency. Thus, for an arbitrary radiating surface, we may write

$$\boxed{\frac{e_v}{a_v} = L_v(T)} \tag{5.15}$$

This statement is known as Kirchhoff's law. Kirchhoff also deduced that within the enclosure the radiation is homogeneous, unpolarized and isotropic. Consider two special cases:

(i) There exist materials which are characterized by $a_v = 0$ for certain frequencies. Since L_v cannot be infinitely large, e_v is also zero for this particular frequency. We conclude: a body can emit only such frequencies which it is capable to absorb.
(ii) If the material of a certain body has the property to absorb completely all radiative energy, then $a_v = 1$. Such a body is called a *blackbody*.

For a blackbody we deduce

$$e_v^b = L_v^b = K_v(T) \tag{5.16}$$

where $K_v(T)$ is known as *Kirchhoff's function*. From (5.14) follows for the emission of an arbitrary body

$$\frac{e_v}{a_v} = \frac{e_v^b}{a_v^b} = e_v^b = K_v(T) \tag{5.17}$$

so that

$$\boxed{e_v = a_v K_v(T)} \tag{5.18}$$

This equation states that the emission of the thermal radiation by an arbitrary body is known if the absorptivity and Kirchhoff's function are known. Finding $K_v(T)$ is equivalent to finding the spectral energy density function $\widehat{u}_v = \widehat{u}_v(T)$ of the radiation field since $\widehat{u}_\nu$ is proportional to L_ν. This becomes evident by recalling that the thermal radiation is of electromagnetic nature whose speed of propagation is $v = c/n$. Here c is the speed of light in vacuum and n the refractive index which may be set equal to 1 for atmospheric problems.

According to (5.1), the energy dE in time dt fills the volume element

$$dV = \mathbf{n} \cdot \boldsymbol{\Omega}\, v\, df\, dt \qquad (5.19)$$

The radiative energy density pertaining to $d\omega$ is then found by dividing (5.1) by (5.19),

$$d\widehat{u} = \frac{dE}{dV} = \frac{L\, d\omega}{v} = \frac{n}{c} L\, d\omega \qquad (5.20)$$

In case of *isotropic radiation*, the integration can be carried out analytically by integrating over the unit sphere with $d\omega = \sin\theta\, d\theta\, d\phi$. The result is

$$\widehat{u} = \frac{nL}{c} \int_0^{2\pi} \int_0^{\pi} \sin\theta\, d\theta\, d\phi = \frac{4\pi n L}{c} \qquad (5.21)$$

or, with reference to frequency

$$\widehat{u}_\nu d\nu = \frac{4\pi n L_\nu}{c} d\nu \qquad (5.22)$$

For the special case of blackbody radiation e_ν^{b} we have

$$\widehat{u}_\nu(T) = \widehat{u}_\nu^{\mathrm{b}}(T) = \frac{4\pi n}{c} K_\nu(T) = \frac{4\pi n}{c} e_\nu^{\mathrm{b}} \qquad (5.23)$$

Therefore, the determination of Kirchhoff's function, expressed by the blackbody emission e_ν^{b}, is equivalent to finding the energy density function $\widehat{u}_\nu(T)$.

5.3 Blackbody radiation

Every physical object emits radiative energy and absorbs a fraction of the incident radiation. If $a_\nu = 1$ for all frequencies, as stated before, we speak of a blackbody. In case that $a_\nu = 0$ for the entire spectrum we call this a *whitebody* or a *perfect mirror*. Both ideal situations can only be realized approximately. Experimentally a blackbody is best realized by a hollow cavity with a small opening permitting the radiation to enter. The radiation in the interior of the cavity is reflected many times. Every reflection is accompanied by absorption resulting finally in nearly perfect absorption. Experimental results on the other hand have shown that the radiation escaping from the opening of a cavity is blackbody radiation.

The electromagnetic radiation field inside the cavity is also known as *cavity radiation* or "*Hohlraumstrahlung*". Since the walls of the cavity continually emit and absorb radiation, *radiative equilibrium* is attained meaning that the characteristics of the radiation do not change any more. This cavity radiation is known to be homogeneous, isotropic and unpolarized. The equilibrium cavity radiation may be described by the temperature of the walls. Experimental evidence shows that the energy density $\widehat{u}_v$ of the cavity radiation does not change if we change the volume of the cavity but hold its temperature constant. Therefore, the energy density is a function of temperature only which may be described by the *caloric equation*

$$E = \widehat{u}(T)V \tag{5.24}$$

If we consider the cavity radiation as a photon gas then, as in the case of an ordinary gas, we must also assign a *pressure to the radiation field*. Thermodynamics does not provide information about the magnitude of this pressure. Electromagnetic theory, however, shows that the pressure exerted by the blackbody radiation is given by

$$p = \frac{\widehat{u}(T)}{3} \tag{5.25}$$

which is the equation of state of the cavity or blackbody radiation. The blackbody radiation is, therefore, completely specified by radiation pressure, the volume and the temperature of the walls assumed to be in equilibrium with the radiation field. This temperature is loosely called the *temperature of the radiation*.

Since the photon gas involves the coordinates p, V and T, it may be treated as a thermodynamic system so that the thermodynamic identities derived in previous chapters can be applied. Then the radiative energy E assumes the role of the *internal energy* and *Gibbs' fundamental equation* (4.7) reads in extensive form

$$T\,dS = dE + p\,dV \tag{5.26}$$

By writing E and the *entropy* S as functions of (V, T) one obtains from (5.26)

$$T\left(\frac{\partial S}{\partial T}\right)_V dT + T\left(\frac{\partial S}{\partial V}\right)_T dV = \left(\frac{\partial E}{\partial T}\right)_V dT + \left(\frac{\partial E}{\partial V}\right)_T dV + p\,dV \tag{5.27}$$

Comparison of the coefficients on both sides of (5.27) yields immediately

$$
\begin{align}
\text{(a)} \quad & \frac{1}{T}\left(\frac{\partial E}{\partial T}\right)_V = \left(\frac{\partial S}{\partial T}\right)_V \\
\text{(b)} \quad & \frac{1}{T}\left[\left(\frac{\partial E}{\partial V}\right)_T + p\right] = \left(\frac{\partial S}{\partial V}\right)_T
\end{align}
\tag{5.28}
$$

Performing partial differentiations of (5.28a) with respect to V and (5.28b) with respect to T we obtain

$$\frac{\partial}{\partial V}\left[\frac{1}{T}\left(\frac{\partial E}{\partial T}\right)_V\right]_T = \frac{\partial^2 S}{\partial V \partial T}$$

$$\frac{\partial}{\partial T}\left[\frac{1}{T}\left(\frac{\partial E}{\partial V}\right)_T + \frac{p}{T}\right]_V = \frac{\partial^2 S}{\partial T \partial V} \implies \tag{5.29}$$

$$\frac{\partial}{\partial V}\left[\frac{1}{T}\left(\frac{\partial E}{\partial T}\right)_V\right]_T = \frac{\partial}{\partial T}\left[\frac{1}{T}\left(\frac{\partial E}{\partial V}\right)_T + \frac{p}{T}\right]_V$$

From (5.29) it may be easily seen that

$$\left(\frac{\partial E}{\partial V}\right)_T = T\left(\frac{\partial p}{\partial T}\right)_V - p \tag{5.30}$$

Substituting for the total radiative energy the expression $E = \widehat{u}(T)V$ we obtain

$$\widehat{u} = \frac{T}{3}\frac{d\widehat{u}}{dT} - \frac{\widehat{u}}{3} \implies T\frac{d\widehat{u}}{dT} = 4\widehat{u} \implies \widehat{u} = bT^4 \tag{5.31}$$

where b is an integration constant. Thus, the radiative energy density is proportional to the fourth power of the temperature. Reference to (5.21) shows the relation between the radiance and the radiation density $\widehat{u}$ so that for $n = 1$

$$L = \frac{cb}{4\pi}T^4 = \frac{\sigma}{\pi}T^4 \quad \text{with} \quad \sigma = \frac{cb}{4} \tag{5.32}$$

This is the *blackbody energy* emitted by a black surface of unit area and unit time confined to the unit solid angle. Equation (5.1) is now used to convert this radiance to the radiation flux F emitted into the hemisphere. First we define the radiation flux (more correctly, this should be called flux density) as

$$dF = \frac{dE}{df\,dt} = \frac{\sigma T^4}{\pi}\cos\theta\,d\omega \tag{5.33}$$

and then integrate over the hemisphere adjacent to the emitting surface. The result is

$$\boxed{F = \frac{\sigma T^4}{\pi}\int_0^{2\pi}\int_0^{\pi/2}\cos\theta\sin\theta\,d\theta\,d\phi = \sigma T^4} \tag{5.34}$$

with $\sigma = 5.67 \times 10^{-8}$ W m^2 K^{-4}. This is the very important *Stefan–Boltzmann law*. This law was first deduced from measurements by *Stefan*. The theory leading to (5.34) was provided by *Boltzmann*.

In passing, we remark that *Planck* has derived an exact expression for $\widehat{u}_\nu$. If this expression is integrated over all frequencies and then converted to flux, the

Stefan–Boltzmann law is obtained again. Therefore, this law is well established by experiment and by two independent theories.

5.4 Entropy of blackbody radiation

With the help of Gibbs' fundamental equation we can calculate the entropy of radiation,

$$dS = \frac{dE + p\,dV}{T} = 4bVT^2\,dT + \frac{4}{3}bT^3\,dV = \frac{4}{3}b\,d(T^3V) \qquad (5.35)$$

and we obtain

$$\boxed{S = \frac{4}{3}bT^3V} \qquad (5.36)$$

According to Planck, there is no constant of integration since $S_0 = S(T_0)$ must vanish as T_0 approaches zero. For $S = $ constant we obtain the equation for an *adiabatic process* within the photon gas,

$$VT^3 = \text{constant} \quad \text{or} \quad pV^{4/3} = \text{constant} \qquad (5.37)$$

since $p = bT^4/3$. Comparison with the corresponding expression for an ideal gas shows that the formal agreement is perfect.

5.5 Summary and additional remarks

In this chapter we have discussed the fact that the energy density of cavity radiation is a function of temperature only. A radiation pressure which is proportional to the energy density may be assigned to the radiation field. This pressure cannot be obtained from thermodynamic reasoning, but it can be found with the help of the electromagnetic theory. The description of the cavity radiation by means of the variables pressure, volume and temperature leads to the idea that cavity radiation may be viewed as a photon gas. Thus it was possible to derive the Stefan–Boltzmann radiation law by aplying the usual laws of thermodynamics. A complete description of the radiation field requires the specification of the energy density as function of temperature and frequency. The methods of thermodynamics are not sufficient to provide the frequency distribution. By employing the methods of quantum statistics it is possible to derive the famous Planck radiation law which describes the spectral energy density as a function of temperature and frequency.

Finally, we wish to remark that at room temperatures the radiation pressure of cavity radiation is quite insignificant. Due to the high temperatures existing in the interior of stars, the radiation pressure is of the same order of magnitude as the pressure of an ideal gas for equally large temperatures.

5.6 Problems

5.1: Consider a radiation field in equilibrium with the cavity walls at temperature T. The photon gas is caused to expand reversibly and isothermally. Find the amount of heat which has to be applied to the walls to keep the wall temperature constant.

5.2: Find the specific heat of constant volume C_v of a radiation field in equilibrium with the cavity walls.

5.3: Show that the Gibbs function for blackbody radiation is zero.

5.4: Calculate C_p for blackbody radiation.

5.5: Do you recognize a problem with the definition of C_p in contrast to C_v?

6

Thermodynamic potentials, identities and stability

In the previous chapters we have partly investigated the three functions e, h and μ. In this chapter we will introduce one more function f which is known as the free energy. These four functions are known as thermodynamic potentials.

6.1 Basic identities

To make things as simple as possible, in the next few paragraphs we will first discuss *one-component systems*. The formal extension to multi-component systems will then become obvious.

The basic law of thermodynamics is *Gibbs' fundamental equation* which combines the first and second law of thermodynamics. For a one-component system we may write

$$T\, ds = de + p\, dv \tag{6.1}$$

We consider the entropy as a function of the internal energy and the specific volume. The entropy $s(v, e)$ is known as the thermodynamic potential s in the variables (v, e). If s is known as function of (v, e) then all state functions can be calculated from it.

The task ahead is to show that important properties of thermodynamic systems can be obtained from the potential $s(v, e)$ and that additional potentials exist which are useful in the treatment of various thermodynamic problems. To continue the discussion we rewrite (6.1) in the form

$$de = T\, ds - p\, dv \tag{6.2}$$

and consider the internal energy as function of (v, s). Expansion of e in these variables gives

$$de = \left(\frac{\partial e}{\partial s}\right)_v ds + \left(\frac{\partial e}{\partial v}\right)_s dv \tag{6.3}$$

Comparison of (6.2) and the exact differential de results in

$$\left(\frac{\partial e}{\partial s}\right)_v = T(v, s), \qquad \left(\frac{\partial e}{\partial v}\right)_s = -p(v, s) \qquad (6.4)$$

In this formula we have explicitly written that p and T are functions of the independent variables (v, s) which are also called the *natural coordinates* of e. The quantities (T, p) are called the *conjugated coordinates* of e. The property that the partial derivatives of e with respect to the independent variables results in other physical quantities led to the name potential (in analogy to the terminology in mechanics) for e and other functions showing the same behavior.

Observing the equality of the mixed partial derivatives of e with respect to v and s results in

$$\frac{\partial^2 e}{\partial v \partial s} = \left(\frac{\partial T}{\partial v}\right)_s, \qquad \frac{\partial^2 e}{\partial s \partial v} = -\left(\frac{\partial p}{\partial s}\right)_v \implies \left(\frac{\partial T}{\partial v}\right)_s = -\left(\frac{\partial p}{\partial s}\right)_v \qquad (6.5a)$$

This is one of the so-called *Maxwell relations* which can also be written in the form

$$\left(\frac{\partial v}{\partial T}\right)_s = -\left(\frac{\partial s}{\partial p}\right)_v \qquad (6.5b)$$

by using the proper relations from theorems on partial derivatives. For details consult the Appendix to this chapter where two important theorems of considerable usefulness are derived.

We will now obtain some additional identities involving the *enthalpy*. Using Gibbs' equation in the form

$$dh = T\, ds + v\, dp \qquad (6.6)$$

and viewing the enthalpy as the potential $h = h(p, s)$ we obtain

$$dh = \left(\frac{\partial h}{\partial s}\right)_p ds + \left(\frac{\partial h}{\partial p}\right)_s dp \qquad (6.7)$$

Comparison of (6.6) with the exact differential (6.7) yields the potential properties

$$\left(\frac{\partial h}{\partial s}\right)_p = T(p, s), \qquad \left(\frac{\partial h}{\partial p}\right)_s = v(p, s) \qquad (6.8)$$

From the equality of the mixed partial derivatives follows another Maxwell relation,

$$\left(\frac{\partial T}{\partial p}\right)_s = \left(\frac{\partial v}{\partial s}\right)_p \qquad (6.9)$$

Additional potentials can be derived by rewriting (6.2)

$$de = d(Ts) - s\, dT - p\, dv \quad \text{or} \quad d(e - Ts) = -s\, dT - p\, dv \qquad (6.10)$$

The new potential

$$f = e - Ts, \qquad (6.11)$$

which he termed the *free energy*, was introduced by *Helmholtz*. In honor of Helmholtz this potential is also known as the *Helmholtz function*. The name "free energy" results from the fact that for isothermal processes the system does not perform work at the expense of the internal energy which remains bound.

We verify this statement with the help of (6.10). Inspection shows that for an isothermal process

$$\int_1^2 d a_{\mathrm{E,iso}} = f_2 - f_1 \qquad (6.12)$$

Suppose that an ideal gas will be compressed isothermally from v_1 to v_2. In this case we can write

$$-\int_1^2 p\, dv = -RT \int_{v_1}^{v_2} \frac{dv}{v} = RT \ln\left(\frac{v_1}{v_2}\right) = f_2 - f_1 > 0, \qquad (6.13)$$

i.e. by doing work on the gas the free energy will be increased while the internal energy remains constant.

In order to obtain additional useful identities, we introduce the definition of the free energy into (6.10),

$$df = -s\, dT - p\, dv \qquad (6.14)$$

By expanding the potential $f(v, T)$ and comparing the result with (6.14), we obtain immediately

$$\left(\frac{\partial f}{\partial T}\right)_v = -s(v, T), \qquad \left(\frac{\partial f}{\partial v}\right)_T = -p(v, T) \qquad (6.15)$$

From the equality of the mixed partial derivatives of f with respect to v and T, we obtain another Maxwell relation,

$$\left(\frac{\partial s}{\partial v}\right)_T = \left(\frac{\partial p}{\partial T}\right)_v \qquad (6.16)$$

Still another and particularly useful potential can be found by rewriting (6.2) as

$$d(e + pv - Ts) = d(h - Ts) = -s\, dT + v\, dp \qquad (6.17)$$

The expression in parentheses was previously identified as the specific *Gibbs function*. Since μ has a form similarly to f, it is also called the *free enthalpy*. Expansion of μ in terms of the independent variables (p, T) gives

$$d\mu = -s\,dT + v\,dp = \left(\frac{\partial \mu}{\partial T}\right)_p dT + \left(\frac{\partial \mu}{\partial p}\right)_T dp \qquad (6.18)$$

From this equation we obtain directly

$$\left(\frac{\partial \mu}{\partial T}\right)_p = -s(p, T), \qquad \left(\frac{\partial \mu}{\partial p}\right)_T = v(p, T) \qquad (6.19)$$

and the Maxwell relation

$$\left(\frac{\partial s}{\partial p}\right)_T = -\left(\frac{\partial v}{\partial T}\right)_p \qquad (6.20)$$

What makes this potential so extremely useful is the fact that the independent variables p and T are part of all standard meteorological observations. For ease of reference, all important identities derived from the potentials are collected in Table 6.1 for the one-component system. In this simple system only two thermodynamic degrees of freedom occur, i.e. its thermodynamic state is completely specified by two independent variables.

In the lower section of Table 6.1 some additional relations are listed. These are needed if the mass concentrations are admitted as additional variables. However, the various relations listed in this table are still valid. All one needs to do to account for a *multi-component system* is to indicate that the concentration variable m^k is held constant in the partial differentiation. We continue the discussion on multi-component systems in later sections of this chapter.

It should be pointed out that the potentials e, h, f, μ are by no means the only potentials in existence. Various other thermodynamic potentials have been defined (*Planck, Massieu*) but they are less useful in our work.

In passing, we would like to remark that v, p and T can also be viewed as potentials. This can be recognized by rewriting (6.2), (6.6) and (6.14) as

$$dv = \frac{T}{p}ds - \frac{1}{p}de \implies v = v(s, e)$$

$$dp = \frac{1}{v}dh - \frac{T}{v}ds \implies p = p(h, s) \qquad (6.21)$$

$$dT = -\frac{1}{s}df - \frac{p}{s}dv \implies T = T(v, f)$$

Table 6.1. *Thermodynamic potentials*

Potential	Natural coordinates	Conjugated coordinates		Maxwell relations
Internal energy	$e(v, s)$,	$de = -p\,dv + T\,ds$	$-p = \left(\dfrac{\partial e}{\partial v}\right)_s,$ $\quad T = \left(\dfrac{\partial e}{\partial s}\right)_v$	$\left(\dfrac{\partial p}{\partial s}\right)_v = -\left(\dfrac{\partial T}{\partial v}\right)_s$
Enthalpy	$h(p, s)$,	$dh = v\,dp + T\,ds$	$v = \left(\dfrac{\partial h}{\partial p}\right)_s,$ $\quad T = \left(\dfrac{\partial h}{\partial s}\right)_p$	$\left(\dfrac{\partial v}{\partial s}\right)_p = \left(\dfrac{\partial T}{\partial p}\right)_s$
Free energy	$f(v, T)$,	$df = -p\,dv - s\,dT$	$-p = \left(\dfrac{\partial f}{\partial v}\right)_T,$ $\quad -s = \left(\dfrac{\partial f}{\partial T}\right)_v$	$\left(\dfrac{\partial p}{\partial T}\right)_v = \left(\dfrac{\partial s}{\partial v}\right)_T$
Free enthalpy	$\mu(p, T)$,	$d\mu = v\,dp - s\,dT$	$v = \left(\dfrac{\partial \mu}{\partial p}\right)_T,$ $\quad -s = \left(\dfrac{\partial \mu}{\partial T}\right)_p$	$\left(\dfrac{\partial v}{\partial T}\right)_p = -\left(\dfrac{\partial s}{\partial p}\right)_T$

$$\left(\frac{\partial \mu}{\partial m^k}\right)_{p,T} = \left(\frac{\partial f}{\partial m^k}\right)_{v,T} = \left(\frac{\partial h}{\partial m^k}\right)_{p,s} = \left(\frac{\partial e}{\partial m^k}\right)_{v,s} = \mu_k - \mu_1, \qquad k = 0, 2, 3$$

Inspection of Table 6.1 shows that no new information is obtained. These potentials already exist in the disguised forms $e = e(v, s)$, $h = h(p, s)$ and $f = f(v, T)$.

6.2 Characteristic properties of thermodynamic potentials

The potentials $e(v, s)$, $h(p, s)$, $f(v, T)$ and $\mu(p, T)$ possess the property that the knowledge of one of the potentials is sufficient to express all characteristic properties of the system in terms of its independent variables. We shall partially demonstrate this for the potential $f(v, T)$. With reference to Table 6.1 we can write the following expression for the internal energy and the enthalpy

$$
\begin{aligned}
e &= f + Ts = f - T \left(\frac{\partial f}{\partial T} \right)_v \\
h &= f + Ts + pv = f - T \left(\frac{\partial f}{\partial T} \right)_v - v \left(\frac{\partial f}{\partial v} \right)_T
\end{aligned}
\tag{6.22}
$$

Hence, e and h are expressed entirely in terms of f and its independent variables. We can now use the relation for the internal energy to express the specific heat capacity c_v as function of f,

$$
c_v = \left(\frac{\partial e}{\partial T} \right)_v = \frac{\partial}{\partial T} \left[f - T \left(\frac{\partial f}{\partial T} \right)_v \right]_v = -T \left(\frac{\partial^2 f}{\partial T^2} \right)_v
\tag{6.23}
$$

It is a somewhat lengthy but by no means difficult exercise to derive an analogous expression for c_p in terms of f and its independent variables. Therefore, we shall refrain from doing this.

We will conclude this paragraph by stating an equation analogous to (6.22) which involves the free enthalpy instead of the free energy. Using the definition of μ and consulting Table 6.1, we can write down immediately

$$
\begin{aligned}
h &= \mu + Ts = \mu - T \left(\frac{\partial \mu}{\partial T} \right)_p \\
e &= \mu + Ts - pv = \mu - T \left(\frac{\partial \mu}{\partial T} \right)_p - p \left(\frac{\partial \mu}{\partial p} \right)_T
\end{aligned}
\tag{6.24}
$$

Equations (6.22) and (6.24) are known as the *Gibbs–Helmholtz equations*. These can be easily extended with the help of Table 6.1. A summary of the Gibbs–Helmholtz equations for all potentials is given in Table 6.2. The first row of this table contains the given potential as function of its natural coordinates while in the following rows the other potentials are derived as function of the given potential.

Table 6.2. Gibbs–Helmholtz relations

	Internal energy	Enthalpy	Free energy	Free enthalpy
$e(v,s)$	—	$h = e - v\left(\frac{\partial e}{\partial v}\right)_s$	$f = e - s\left(\frac{\partial e}{\partial s}\right)_v$	$\mu = e - v\left(\frac{\partial e}{\partial v}\right)_s - s\left(\frac{\partial e}{\partial s}\right)_v$
$h(p,s)$	$e = h - p\left(\frac{\partial h}{\partial p}\right)_s$	—	$f = h - p\left(\frac{\partial h}{\partial p}\right)_s - s\left(\frac{\partial h}{\partial s}\right)_p$	$\mu = h - s\left(\frac{\partial h}{\partial s}\right)_p$
$f(v,T)$	$e = f - T\left(\frac{\partial f}{\partial T}\right)_v$	$h = f - v\left(\frac{\partial f}{\partial v}\right)_T - T\left(\frac{\partial f}{\partial T}\right)_v$	—	$\mu = f - v\left(\frac{\partial f}{\partial v}\right)_T$
$\mu(p,T)$	$e = \mu - p\left(\frac{\partial \mu}{\partial p}\right)_T - T\left(\frac{\partial \mu}{\partial T}\right)_p$	$h = \mu - T\left(\frac{\partial \mu}{\partial T}\right)_p$	$f = \mu - p\left(\frac{\partial \mu}{\partial p}\right)_T$	—

6.3 Equilibrium conditions and thermodynamic stability

6.3.1 Equilibrium conditions

We have already pointed out that the thermodynamic potentials $S(V, E)$, $E(V, S)$, $F(V, T)$, $H(p, S)$ and $G(p, T)$ possess a very significant property. If one of these potentials is known, then all characteristic thermodynamic properties of the system can be expressed from this potential. In a later chapter we will derive analytical expressions for these potentials for the atmospheric system describing the cloudy air. Presently, however, we wish to show that the potentials can also be used to describe the thermodynamic equilibrium. We will demonstrate this for the simple situation with two degrees of freedom.

We begin by restating Gibbs' fundamental equation for the one-component system

$$T \, dS \geq dE + p \, dV \tag{6.25}$$

The equal sign refers to the reversible process, the larger sign to the irreversible (natural) process. The question arises under which conditions the equilibrium is attainable. Additional equilibrium conditions will be obtained by replacing the internal energy in (6.25) by the free energy and the free enthalpy. Direct substitution of these potentials yields

$$
\begin{aligned}
\text{(a)} \quad & T \, dS \geq d(F + TS) + p \, dV \\
\text{(b)} \quad & T \, dS \geq d(G + TS - pV) + p \, dV
\end{aligned}
\tag{6.26}
$$

or

$$
\begin{aligned}
\text{(a)} \quad & dF \leq -S \, dT - p \, dV \\
\text{(b)} \quad & dG \leq -S \, dT + V \, dp
\end{aligned}
\tag{6.27}
$$

We will consider various important cases.

6.3.1.1 The isolated system

Such a system is characterized by the restrictive conditions: $E = \text{constant}, dE = 0$, $V = \text{constant}, dV = 0$. From (6.25) follows directly $dS \geq 0$. To verify this relation we simply have to recall that in an isolated system the entropy must increase until it reaches a maximum value. Thus, the equilibrium condition of the isolated system reads $S = S_{\text{max}}, dS = 0$.

6.3.1.2 The isothermal–isochoric system

In this system the restrictive conditions are given by $T = \text{constant}, dT = 0$, $V = \text{constant}, dV = 0$ so that according to (6.27a) $dF \leq 0$. Therefore, isothermal and isochoric processes reach equilibrium whenever the free energy has a minimum value, i.e. $F = F_{min}$, $dF = 0$.

6.3.1.3 The isothermal–isobaric system

Now we have the restrictive conditions $T = $ constant, $dT = 0$, $p = $ constant, $dp = 0$. With these requirements we obtain from (6.27b) $dG \leq 0$. For the equilibrium we have $G = G_{min}$, $dG = 0$.

6.3.2 Thermodynamic stability

Before we discuss the concept of the thermodynamic stability of systems, we wish to refer to the stability concept of mechanics. If a mechanical system in equilibrium is given a small displacement, it may or may not return to the original state. If the system tends to return to equilibrium, given a sufficiently small displacement, the equilibrium is stable, otherwise the equilibrium is unstable. If the system has no tendency to move either toward or away from equilibrium, the equilibrium is said to be neutral or indifferent. A ball placed (1) at the bottom of a spherical bowl, (2) on top of a spherical cap, and (3) on a plane horizontal surface are examples of stable, unstable and neutral equilibrium, respectively.

The stability concept of mechanics can also be applied to thermodynamic systems. We think of some thermodynamic system to be infinitesimally disturbed by an intervention from the external surroundings so that the system is slightly displaced from equilibrium. Now we imagine some suitable constraint to confine the system to a new equilibrium state. If the constraint is removed and the system is left to itself, then one of the following processes will occur.

(i)   The system returns (without intervention) to the original state. This is *stable thermodynamic equilibrium.*

(ii) The system continues the displacement from the original equilibrium state until it reaches a new equilibrium state. The system is *unstable.*

(iii) The system remains in its new equilibrium state. The original state is called *neutral.*

Examples of unstable states are overheated water ($p = 1013.25$ hPa, $T = 378$ K) and undercooled water ($p = 1013.25$ hPa, $T = 263$ K). If the overheated water or the undercooled water is brought in contact with the corresponding (real gas) stable vapor or ice phase, they will rapidly transform to this phase.

If an equilibrium state is stable to a finite displacement, the system is said to be *absolutely stable.* We are now ready to work out the stability criteria for thermodynamic equilibrium. For simplicity we consider an isolated system ($dE = 0, dV = 0$). In case of a stable equilibrium state, the system will return to its original equilibrium state if its variables of state are changed infinitesimally. For the system to be stable, the entropy must not only be characterized by stationarity ($dS = 0$) but also by $S = S_{max}$.

Before we formulate the stability concept mathematically, we imagine an isolated system which is subdivided into two equally large partial systems, each having the same internal energy. The entropy of the complete system is then given by

$$S(2E, 2V) = 2S(E, V) \tag{6.28}$$

We now vary the first part of the system by (dE, dV) and the second part by $(-dE, -dV)$. This variation in the state variables does not change the internal energy and the volume of the total system, as required. To satisfy the requirement that the entropy of the original state has reached a maximum value, we must have

$$2S(E, V) > S(E + dE, V + dV) + S(E - dE, V - dV) \tag{6.29}$$

The Taylor expansion of the right-hand side immediately yields

$$\left(\frac{\partial^2 S}{\partial E^2}\right)_V (dE)^2 + 2\frac{\partial^2 S}{\partial E \partial V}dE\,dV + \left(\frac{\partial^2 S}{\partial V^2}\right)_E (dV)^2 < 0 \tag{6.30}$$

where third and higher order terms are neglected. The maximum condition requires that (6.30) be less than zero. We will now rewrite (6.30) to bring it into a form which lends itself more readily to physical interpretation. From Gibbs' equation we recognize that $S = S(E, V)$. Writing down the exact differential of S and comparing this expression with Gibbs' equation, we immediately obtain

$$\left(\frac{\partial S}{\partial E}\right)_V = \frac{1}{T}, \qquad \left(\frac{\partial S}{\partial V}\right)_E = \frac{p}{T} \tag{6.31}$$

The differentials of (6.31) will be needed shortly. They are given by

$$d\left(\frac{1}{T}\right) = \left(\frac{\partial^2 S}{\partial E^2}\right)_V dE + \left(\frac{\partial^2 S}{\partial E \partial V}\right) dV \tag{6.32}$$

$$d\left(\frac{p}{T}\right) = \left(\frac{\partial^2 S}{\partial E \partial V}\right) dE + \left(\frac{\partial^2 S}{\partial V^2}\right)_E dV \tag{6.33}$$

Multiplying (6.32) by dE and (6.33) by dV and adding the results gives precisely the left-hand side of (6.30) or

$$d\left(\frac{1}{T}\right) dE + d\left(\frac{p}{T}\right) dV < 0 \tag{6.34}$$

Carrying out the differential operation we obtain at once

$$\left(\frac{dE}{T} + \frac{p}{T}dV\right)dT - dp\,dV > 0 \tag{6.35}$$

The expression in parentheses is part of Gibbs' fundamental equation so that

$$dS\,dT - dp\,dV > 0 \tag{6.36}$$

We summarize: a thermodynamic system is found to be in a stable equilibrium whenever the inequality (6.34) or (6.36) is valid for arbitrary infinitesimal changes of the variables of state. Next we will give two examples on stability conditions.

Example 1: Condition on the heat capacity at constant volume

Application of (6.34) to isochoric processes results in

$$d\left(\frac{1}{T}\right)dE < 0 \tag{6.37}$$

Ignoring viscous processes the heat differential $đQ$ in the first law can always be written as

$$đQ = Mc_v\,dT = dE + p\,dV \tag{6.38}$$

where c_v is the specific heat capacity at constant volume. For an isochoric process we may write

$$dE = Mc_v\,dT \tag{6.39}$$

so that (6.37) can be written as

$$-\frac{1}{T^2}dT\,Mc_v\,dT < 0 \tag{6.40}$$

Using the definition (3.55) we obtain

$$Mc_v = \left(\frac{\partial E}{\partial T}\right)_V > 0 \tag{6.41}$$

In words: in a stable thermodynamic system the heat capacity at constant volume is always positive. This intuitively obvious result, therefore, has a firm theoretical background.

Example 2: Condition on the isothermal compressibility

The *isothermal compressibility*, needed later, is defined by

$$\chi = -\frac{1}{V}\left(\frac{\partial V}{\partial p}\right)_T \tag{6.42}$$

Dividing (6.36) by $V(dp)^2$ and using the isothermal condition $dT = 0$, we obtain

$$\chi = -\frac{1}{V}\left(\frac{\partial V}{\partial p}\right)_T > 0 \qquad (6.43)$$

In words: in a stable thermodynamic system the isothermal compressibility is always positive. Again, an intuitively obvious result has a firm theoretical background.

The stability considerations can be extended to the more complicated multi-component systems. The reader who is interested in this subject may consult a more specialized book, for example, Reichel (1980).

6.4 Identities of multi-component systems

6.4.1 Variational notation

In atmospheric systems the quantity $\psi(x^i)$ with $x^i = p, T, m^0, m^2, m^3$ (m^1 has been eliminated) in general depends on the spatial variables q^k and on time t. Thus, we may write $x^i = x^i(q^k, t)$ where q^k is a variable of space. Now viewing ψ as a function of space coordinates and time, we can write various differential expressions:

$$\frac{\partial \psi}{\partial t} = \frac{\partial \psi}{\partial x^n}\frac{\partial x^n}{\partial t}, \qquad \frac{d\psi}{dt} = \frac{\partial \psi}{\partial x^n}\frac{dx^n}{dt}, \qquad \frac{\partial \psi}{\partial q^k} = \frac{\partial \psi}{\partial x^n}\frac{\partial x^n}{\partial q^k} \qquad (6.44)$$

In the last expression we identify in succession $q^k = x^1, x^2, x^3$ and multiply the terms $\partial \psi/\partial x^k$ by the corresponding unit vectors $\mathbf{i}_k$. Adding the three resulting expressions yields the important equation

$$\nabla \psi = \frac{\partial \psi}{\partial x^n}\nabla x^n \qquad (6.45)$$

In order to use a very compact notation, we will introduce for these differential expressions the variational symbol $\delta = d, d/dt, \partial/\partial t, \partial/\partial q^k, \nabla, \ldots$ so that for (6.44) and (6.45) we have the common notation

$$\delta \psi = \frac{\partial \psi}{\partial x^n}\delta x^n \qquad (6.46)$$

Since the budget operator is not an ordinary differential expression, we recognize that $\delta \neq D/DT$. Moreover, the inexact differential operators dq or and dw cannot be replaced by δq or δw since neither q nor w are functions of state.

We will now apply the variational notation to the important equation (1.20) and identify the general function ψ as the chemical potential μ. Since m^1 was eliminated we may write

$$\delta\mu = \left(\frac{\partial\mu}{\partial p}\right)_{T,m^k}\delta p + \left(\frac{\partial\mu}{\partial T}\right)_{p,m^k}\delta T + \sum_{k=0}^{3}(\mu_k - \mu_1)\delta m^k \qquad (6.47)$$

For comparison purposes we need to state another form of this equation. To accomplish this, we apply the methods of Section 6.1 by which we obtained various thermodynamic identities. We will now generalize the equation leading to the identity (6.19) by retaining the concentration term. We start with the variational form of (4.13)

$$\delta\mu = v\delta p - s\delta T + \left(\frac{\partial\mu}{\partial m^0}\right)_{p,T}\delta m^0 + \left(\frac{\partial\mu}{\partial m^2}\right)_{p,T}\delta m^2 + \left(\frac{\partial\mu}{\partial m^3}\right)_{p,T}\delta m^3 \quad (6.48)$$

where the concentration m^1 has been eliminated. Comparison of (6.47) and (6.48) results in the identities

$$\left(\frac{\partial\mu}{\partial p}\right)_{T,m^k} = v, \qquad \left(\frac{\partial\mu}{\partial T}\right)_{p,m^k} = -s, \qquad \left(\frac{\partial\mu}{\partial m^k}\right)_{p,T} = \mu_k - \mu_1 \qquad (6.49)$$

Reference to Table 6.1 shows that the form of the identities for s and v, as expected, have not changed. To account for the multi-component system, we simply have to hold the concentrations m^k constant in the partial differentiations. The same is true for all other identities in Table 6.1 which were obtained from Gibbs' fundamental equation for the one-component system. The relation $(\partial\mu/\partial m^k)_{p,T}$ is not new either since we could have obtained it directly from (1.21) with $\psi = \mu$.

By subjecting (6.47) to Legendre transformations to find δf, δh and δe we can easily show that

$$\left(\frac{\partial\mu}{\partial m^k}\right)_{p,T} = \left(\frac{\partial f}{\partial m^k}\right)_{v,T} = \left(\frac{\partial h}{\partial m^k}\right)_{p,s} = \left(\frac{\partial e}{\partial m^k}\right)_{v,s} = \mu_k - \mu_1, \qquad k = 0, 2, 3$$
$$(6.50)$$

We have already added these identities to Table 6.1.

For ease of reference we will now collect various variational statements involving the potentials.

(a)	$e = e(v, s, m^0, m^2, m^3)$,	$\delta e = T\delta s - p\delta v + \mu_n \delta m^n$
(b)	$h = h(p, s, m^0, m^2, m^3)$,	$\delta h = T\delta s + v\delta p + \mu_n \delta m^n$
(c)	$f = f(v, T, m^0, m^2, m^3)$,	$\delta f = -s\,\delta T - p\delta v + \mu_n \delta m^n$
(d)	$\mu = \mu(p, T, m^0, m^2, m^3)$,	$\delta \mu = -s\,\delta T + v\delta p + \mu_n \delta m^n$

$$(e) \quad \mu_n \delta m^n = \sum_{k=0}^{3} (\mu_k - \mu_1)\delta m^k$$

$$(6.51)$$

6.4.2 Kelvin–De Donder and Gibbs–Duhem relations

Thermodynamics is characterized by innumerable identities. Only a few of these have been derived so far which are particularly important for atmospheric systems. In the concluding section of this chapter we will derive some additional thermodynamic identities which will also be of great usefulness.

The starting point in the derivation, as usual, is *Gibbs' fundamental equation* (6.51b)

$$\delta s = \frac{1}{T}\delta h - \frac{v}{T}\delta p - \frac{\mu_n}{T}\delta m^n \tag{6.52}$$

Replacing $\delta h - v\delta p$ according to (3.41) and using the definitions for c_p and γ as stated in (3.43), we obtain

$$\begin{aligned}
\delta s &= \frac{1}{T}[c_p\delta T + \gamma\delta p + (h_n - \mu_n)\delta m^n] \\
&= \frac{1}{T}\left(c_p\delta T + \gamma\delta p + \sum_{k=0}^{3}[(h_k - h_1) - (\mu_k - \mu_1)]\delta m^k\right)
\end{aligned} \tag{6.53}$$

Since s is a variable of state, we may write an equivalent expression for δs using (1.20) and (1.21)

$$\delta s = \left(\frac{\partial s}{\partial T}\right)_{p,m^k}\delta T + \left(\frac{\partial s}{\partial p}\right)_{T,m^k}\delta p + \sum_{k=0}^{3}(s_k - s_1)\delta m^k \tag{6.54}$$

Comparison of coefficients results in

$$\begin{aligned}
s_k - s_1 &= \frac{1}{T}[(h_k - h_1) - (\mu_k - \mu_1)], && k = 0, 2, 3 \\
s_k - s_1 &= -\frac{l_{k1} + a_{k1}}{T}, && k = 2, 3
\end{aligned} \tag{6.55}$$

$$\left(\frac{\partial s}{\partial T}\right)_{p,m^k} = \frac{c_p}{T}, \qquad \left(\frac{\partial s}{\partial p}\right)_{T,m^k} = \frac{\gamma}{T}$$

where use has been made of (3.43) and (4.23).

Additional important identities are found from the equality of the mixed partial derivatives of the specific entropy,

$$
\frac{\partial^2 s}{\partial T \partial p} = \frac{\partial^2 s}{\partial p \partial T} \implies \frac{\partial}{\partial T}\left(\frac{\gamma}{T}\right)_{p,m^k} = \frac{\partial}{\partial p}\left(\frac{c_p}{T}\right)_{T,m^k}
$$

$$
\frac{\partial^2 s}{\partial m^k \partial p} = \frac{\partial^2 s}{\partial p \partial m^k} \implies \frac{\partial}{\partial m^k}\left(\frac{\gamma}{T}\right)_{p,T} = -\frac{\partial}{\partial p}\left(\frac{l_{k1}+a_{k1}}{T}\right)_{T,m^k}, \quad k=2,3
$$

$$
\frac{\partial^2 s}{\partial m^k \partial T} = \frac{\partial^2 s}{\partial T \partial m^k} \implies \frac{\partial}{\partial m^k}\left(\frac{c_p}{T}\right)_{p,T} = -\frac{\partial}{\partial T}\left(\frac{l_{k1}+a_{k1}}{T}\right)_{p,m^k}, \quad k=2,3
$$

$$(6.56)$$

Carrying out the differentiations, we immediately obtain

$$
-\frac{\gamma}{T} + \left(\frac{\partial \gamma}{\partial T}\right)_{p,m^k} = \left(\frac{\partial c_p}{\partial p}\right)_{T,m^k}
$$

$$
\left(\frac{\partial \gamma}{\partial m^k}\right)_{p,T} = -\left(\frac{\partial(l_{k1}+a_{k1})}{\partial p}\right)_{T,m^k}, \quad k=2,3 \quad (6.57)
$$

$$
\left(\frac{\partial c_p}{\partial m^k}\right)_{p,T} = \frac{l_{k1}}{T} - \left(\frac{\partial l_{k1}}{\partial T}\right)_{p,m^k} - T\frac{\partial}{\partial T}\left(\frac{a_{k1}}{T}\right)_{p,m^k}, \quad k=2,3
$$

For atmospheric systems γ is given by $-v$ in good approximation. These equations will now be combined with the *Clausius–Kirchhoff relations* (see Section 3.4) to give some very handy identities which will be needed later when we investigate the dependency of the saturation vapor pressure on the state variables. We proceed as follows. The expressions $(\partial c_p/\partial p)_{T,m^k}$, $(\partial c_p/\partial m^k)_{p,T}$ and $(\partial \gamma/\partial m^k)_{p,T}$ will be eliminated since they appear in (6.57) and in the Clausius–Kirchhoff relations. The result, known as the Kelvin–De Donder relations, is given by

$$
\begin{array}{ll}
\text{(a)} & \left(\dfrac{\partial v}{\partial T}\right)_{p,m^k} = -\dfrac{\gamma}{T} \approx \dfrac{v}{T} \\[3mm]
\text{(b)} & l_{k1} = T^2 \dfrac{\partial}{\partial T}\left(\dfrac{a_{k1}}{T}\right)_{p,m^k}, \quad k=2,3 \\[3mm]
\text{(c)} & \left(\dfrac{\partial v}{\partial m^k}\right)_{p,T} = \left(\dfrac{\partial a_{k1}}{\partial p}\right)_{T,m^k} = v_k - v_1, \quad k=2,3
\end{array}
$$

$$(6.58)$$

We conclude this section by briefly deriving two additional identities known as the Gibbs–Duhem relations which will also be needed later. The starting point in the derivation is the equation (1.19) in variational notation

$$
\left(\frac{\partial \psi}{\partial p}\right)_{T,m^k}\delta p + \left(\frac{\partial \psi}{\partial T}\right)_{p,m^k}\delta T = m^n \delta \psi_n \quad (6.59)
$$

With $\psi = \mu$ and using the proper identities of Table 6.1, this equation may be changed to the form

$$\boxed{v\,\delta p - s\,\delta T = m^n \delta \mu_n} \tag{6.60}$$

This is the *entropy form of the Gibbs–Duhem equation*. We transform this equation to the energy form by first multiplying (6.60) by $1/T$ and then include $1/T$ in the differentials. By doing so we arrive at the expression $(Ts - pv + \mu)$ which is equivalent to the internal energy e. The result is

$$\boxed{e\,\delta\left(\frac{1}{T}\right) + v\,\delta\left(\frac{p}{T}\right) = m^n \delta\left(\frac{\mu_n}{T}\right)} \tag{6.61}$$

6.5 Summary

In this chapter we have introduced the free energy f. The four functions e, h, f and μ are known as thermodynamic potentials. We have used these potentials to derive an array of thermodynamic identities which will prove useful in the description of thermodynamic systems. Moreover, we have discussed thermodynamic equilibrium conditions in terms of these potentials and the concept of thermodynamic stability. In the last part of this chapter we have introduced the variational δ to represent various differential operators. By using this type of notation we will save some work otherwise required in the derivation of various equations. Two important theorems on partial derivatives will be stated in the Appendix to this chapter.

6.6 Appendix

There are two simple theorems that are needed quite frequently to relate partial derivatives. We will derive these very briefly. Suppose there exists a relation among the three thermodynamic variables x, y, z in the form

$$f(x, y, z) = 0 \tag{6.62}$$

We can think of (6.62) of being solved in the form $x = x(y, z)$ or $y = y(x, z)$. The differentials of these functions are given by

$$\text{(a)} \quad dx = \left(\frac{\partial x}{\partial y}\right)_z dy + \left(\frac{\partial x}{\partial z}\right)_y dz$$

$$\text{(b)} \quad dy = \left(\frac{\partial y}{\partial x}\right)_z dx + \left(\frac{\partial y}{\partial z}\right)_x dz \tag{6.63}$$

Substituting (6.63b) in (6.63a) gives

$$dx = \left(\frac{\partial x}{\partial y}\right)_z \left[\left(\frac{\partial y}{\partial x}\right)_z dx + \left(\frac{\partial y}{\partial z}\right)_x dz\right] + \left(\frac{\partial x}{\partial z}\right)_y dz \tag{6.64}$$

and rearranging results in

$$dx = \left(\frac{\partial x}{\partial y}\right)_z \left(\frac{\partial y}{\partial x}\right)_z dx + \left[\left(\frac{\partial x}{\partial y}\right)_z \left(\frac{\partial y}{\partial z}\right)_x + \left(\frac{\partial x}{\partial z}\right)_y\right] dz \qquad (6.65)$$

Only two of the three coordinates x, y, z are independent, say x and z. Then (6.65) must be true for all dx and dz. If $dx \neq 0$ and $dz = 0$, we have

$$\left(\frac{\partial x}{\partial y}\right)_z \left(\frac{\partial y}{\partial x}\right)_z = 1 \implies \left(\frac{\partial x}{\partial y}\right)_z = \frac{1}{\left(\frac{\partial y}{\partial x}\right)_z} \qquad (6.66)$$

If $dx = 0$ and $dz \neq 0$ then

$$\left(\frac{\partial x}{\partial y}\right)_z \left(\frac{\partial y}{\partial z}\right)_x + \left(\frac{\partial x}{\partial z}\right)_y = 0 \qquad (6.67)$$

Using (6.66) the equation (6.67) can be written as

$$\left(\frac{\partial x}{\partial y}\right)_z \left(\frac{\partial y}{\partial z}\right)_x \left(\frac{\partial z}{\partial x}\right)_y = -1 \qquad (6.68)$$

We will apply this equation to the ideal gas with $x = v$, $y = p$, $z = T$.

$$\left(\frac{\partial v}{\partial p}\right)_T = -\frac{v}{p}, \qquad \left(\frac{\partial T}{\partial v}\right)_p = \frac{T}{v}, \qquad \left(\frac{\partial p}{\partial T}\right)_v = \frac{p}{T}$$

$$\left(\frac{\partial v}{\partial p}\right)_T \left(\frac{\partial T}{\partial v}\right)_p \left(\frac{\partial p}{\partial T}\right)_v = -1 \qquad (6.69)$$

6.7 Problems

6.1: Verify equation (6.50).

6.2: Show that

$$\left(\frac{\partial e}{\partial v}\right)_T = T\left(\frac{\partial p}{\partial T}\right)_v - p, \qquad \left(\frac{\partial h}{\partial p}\right)_T = -T\left(\frac{\partial v}{\partial T}\right)_p + v$$

6.3: Let δ represent a small variation. State the equilibrium and minimum conditions for (6.27). Justify your results.

6.4: Prove the validity of the following expressions:

$$h = \left(\frac{\partial e/v}{\partial 1/v}\right)_s, \qquad\qquad e = \left(\frac{\partial h/p}{\partial 1/p}\right)_s$$

$$f = T\int e(v, T)d\left(\frac{1}{T}\right), \qquad \mu = T\int h(p, T)d\left(\frac{1}{T}\right)$$

6.5: Show that

$$\frac{c_p}{c_v} = \left(\frac{\partial s}{\partial T}\right)_p\left(\frac{\partial T}{\partial s}\right)_v, \qquad c_p - c_v = T\left(\frac{\partial p}{\partial T}\right)_v\left(\frac{\partial v}{\partial T}\right)_p$$

6.6: The Helmholtz or free energy function may be expressed as $f = -RT\ln Z$ where $Z = Z(v, T)$. Find expressions for p, s, e, h, μ in terms of (v, T, Z).

6.7: Verify the following expressions:

$$e = -T^2\left(\frac{\partial f/T}{\partial T}\right)_v, \qquad \left(\frac{\partial e}{\partial v}\right)_T = T^2\left(\frac{\partial p/T}{\partial T}\right)_v, \qquad \left(\frac{\partial e}{\partial s}\right)_T = -p^2\left(\frac{\partial p/T}{\partial p}\right)_v$$

7

The constitutive equations for irreversible fluxes

In order to evaluate the prognostic form of the first law of thermodynamics, we must be able to express the as yet undetermined fluxes in terms of suitable thermodynamic variables and the so-called phenomenological coefficients. To accomplish this task, we make use of a theory first developed by Onsager (1931) and then amplified by Eckart (1940), Casimir (1945) and by others.

Before we state this theory in the form applicable to the complicated atmospheric processes, we will first study two illuminating examples. These two examples are too simple to model atmospheric processes, but they will give considerable insight into the structure of the theory. Frictional effects will be ignored. Our reference goes to an excellent article by Keller which is part of the outstanding book on thermodynamics by Päsler (1975).

7.1 Entropy production due to interior heat and mass exchange

We consider two containers as shown in Figure 7.1. Each of these contains the same one-component fluid. The system is isolated with respect to the surroundings.

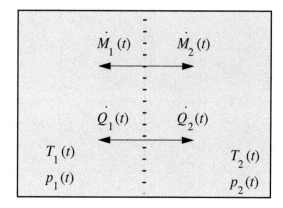

Fig. 7.1 Heat and mass exchange within an isolated system.

The two containers are separated by a porous wall so that heat and mass may be transported from one part of the system to the other. The state variables p and T characterizing the system will be expressed as functions of time. To avoid the involvement of the spatial variables, we assume that the mass and heat flow through the wall is slow in comparison to the equalization processes within the individual containers. Moreover, we postulate that the difference in the magnitude of the variables between the two containers, the partial systems, is small at all times. This condition is stated by

$$\frac{|\psi_1(t) - \psi_2(t)|}{\max[\psi_1(t), \psi_2(t)]} \ll 1 \quad \text{with} \quad \psi = p, T \qquad (7.1)$$

The thermodynamic state of each subsystem i, $i = 1, 2$ is completely determined by the internal energy E_i, the mass M_i and the constant volume V_i of the partial systems. The intensive quantities p_i, T_i and μ_i can be formally determined from *Gibbs' fundamental equation*

$$dS(E_i, M_i, V_i) = \frac{1}{T_i}(dE_i + p_i dV_i - \mu_i dM_i), \qquad i = 1, 2 \qquad (7.2)$$

Expanding dS as an exact differential gives

$$dS(E_i, M_i, V_i) = \left(\frac{\partial S}{\partial E_i}\right)_{M_i, V_i} dE_i + \left(\frac{\partial S}{\partial M_i}\right)_{E_i, V_i} dM_i$$
$$+ \left(\frac{\partial S}{\partial V_i}\right)_{E_i, M_i} dV_i, \quad i = 1, 2 \qquad (7.3)$$

Comparison of coefficients results in

$$\left(\frac{\partial S}{\partial E_i}\right)_{M_i, V_i} = \frac{1}{T_i}, \quad \left(\frac{\partial S}{\partial M_i}\right)_{E_i, V_i} = -\frac{\mu_i}{T_i}, \quad \left(\frac{\partial S}{\partial V_i}\right)_{E_i, M_i} = \frac{p_i}{T_i}, \quad i = 1, 2 \qquad (7.4)$$

These identities are already known to us.

Before continuing, a few general remarks on open systems will be helpful. Let us consider a constant volume fluid system exchanging heat and mass with its surroundings. The differential change of the internal energy of the system can be decomposed. One part, denoted by dE_M, is associated with the mass flow while the remaining part is due to the immaterial heat flow $đQ$. If we add the mass dM to the system, the internal energy will increase due to the mass flow by the amount $e\,dM$ and due to the work which is required to introduce the mass into the system against the pressure existing there. Therefore, the change of the internal energy dE_M associated with dM is given by

$$dE_M = e\,dM - p\,dV = (e + pv)dM = h\,dM \qquad (7.5)$$

where $-dV = v\,dM$ represents the volume occupied by dM. The total internal energy change is then given by

$$dE = dE_M + đ Q = h\,dM + đ Q \tag{7.6}$$

Substituting (7.6) into Gibbs' fundamental equation and recalling that the volume of the system is constant, we obtain

$$dS = \frac{1}{T}(dE - \mu\,dM) = \frac{1}{T}(e + pv - \mu)dM + \frac{đ Q}{T} = s\,dM + \frac{đ Q}{T} \tag{7.7}$$

since $\mu = e + pv - Ts$. Thus the total entropy change of the system consists of two contributions, that is, the addition of mass and heat.

Application of (7.6) and (7.7) to the two partial systems of the above example gives

$$
\begin{aligned}
\text{(a)} \quad & \frac{dE_i}{dt} = h_i\frac{dM_i}{dt} + \frac{đ Q_i}{dt} && \text{or} \quad \dot{E}_i = h_i\dot{M}_i + \dot{Q}_i, && i = 1, 2 \\
\text{(b)} \quad & \frac{dS_i}{dt} = s_i\frac{dM_i}{dt} + \frac{1}{T_i}\frac{đ Q_i}{dt} && \text{or} \quad \dot{S}_i = s_i\dot{M}_i + \frac{\dot{Q}_i}{T_i}, && i = 1, 2
\end{aligned}
\tag{7.8}
$$

Next we formulate the conservation laws for mass and heat of the total system. Conservation of mass gives

$$
\begin{aligned}
\text{(a)} \quad & M_1 + M_2 = \text{constant} \\
\text{(b)} \quad & \dot{M}_1 = -\dot{M}_2 = \dot{M}
\end{aligned}
\tag{7.9}
$$

so that the two mass transports are of equal magnitude but of opposite direction. Since the system is isolated, conservation of energy gives

$$
\begin{aligned}
\text{(a)} \quad & dE = dE_1 + dE_2 = 0 \\
\text{(b)} \quad & \dot{E}_1 = -\dot{E}_2
\end{aligned}
\tag{7.10}
$$

For the immaterial heat transport between the two partial systems we write

$$\dot{Q}_1 = \dot{Q} \tag{7.11}$$

Combination of (7.8)–(7.11) yields

$$\dot{Q}_2 = -\dot{Q} + (h_2 - h_1)\dot{M} \tag{7.12}$$

This equation shows that the heat transport between the partial systems is not equal and of opposite sign. This is the case only if $\dot{M} = 0$ or if $h_1 = h_2$.

The first law has not given us any information about the magnitude of $\dot{Q}$ and $\dot{M}$. This is not surprising since heat and mass transport depend on the properties of the wall. Therefore, it is necessary to include additional information.

If the heat and mass transport between the partial systems takes place for a long period of time ($t \to \infty$), then equilibrium occurs.

$$p_1(t \to \infty) = p_2(t \to \infty) \quad \text{or} \quad p(E_{1,\infty}, V_1, M_{1,\infty}) = p(E_{2,\infty}, V_2, M_{2,\infty})$$
$$T_1(t \to \infty) = T_2(t \to \infty) \quad \text{or} \quad T(E_{1,\infty}, V_1, M_{1,\infty}) = T(E_{2,\infty}, V_2, M_{2,\infty})$$
$$(7.13)$$

Integration of (7.9b) and (7.10b) between the initial ($t = 0$) and final states ($t = \infty$) of the system then gives

$$M_{1,\infty} + M_{2,\infty} = M_{1,0} + M_{2,0}, \qquad E_{1,\infty} + E_{2,\infty} = E_{1,0} + E_{2,0} \qquad (7.14)$$

Now four equations (7.13), (7.14) are at our disposal to determine the four unknown parameters of the asymptotic equilibrium states $M_{1,\infty}$, $M_{2,\infty}$, $E_{1,\infty}$ and $E_{2,\infty}$.

To complete the transport problem we need to set up proper conditions which are in harmony with observations. These show that heat and mass fluxes occur between the partial systems as long as temperature and pressure differences or both of these exist. If pressure and temperature differences do not exist then the system is in thermodynamic equilibrium so that neither heat nor mass transport takes place. Observations point out that certain relations exist between these fluxes. To get a better understanding of such relations we will first consider some special cases.

7.1.1 Heat conduction only

In this case the wall must be nonporous. Due to the temperature difference a heat flow takes place as expressed by the well-known *Fourier law*,

$$\dot{Q} = \frac{F}{d} l_q (T_2 - T_1), \qquad l_q > 0 \qquad (7.15)$$

where $l_q > 0$ is the *heat conduction coefficient* which certainly depends on the properties of the wall. The quantities F and d represent here and in the following formulas the cross-section and the thickness of the wall.

7.1.2 Isothermal permeation

Both systems have the same temperature but the pressure differs. According to *Hagen*, the mass flux is given by a transport law analogous to equation (7.15),

$$\dot{M} = \frac{F}{d} l_m (p_2 - p_1), \qquad l_m > 0 \qquad (7.16)$$

where $l_m > 0$ is the thermodynamic *permeability coefficient* which certainly depends on the properties of the wall.

7.1.3 Thermal osmosis

In both partial systems the pressure is the same but the temperature differs. There exists a mass flow, known as *thermal osmosis*, which is described by

$$\dot{M} = \frac{F}{d} l_{mq}(T_2 - T_1) \tag{7.17}$$

The coefficient l_{mq} can be either positive or negative so that the mass flux can go in either direction.

7.1.4 Osmotic thermal flow

Both systems have the same temperature but the pressure differs. A heat flux takes place which is given by

$$\dot{Q} = \frac{F}{d} l_{qm}(p_2 - p_1) \tag{7.18}$$

where the coefficient l_{qm} can be either positive or negative. This equation states that a pressure difference may cause a heat flux. Thus, the initial thermal equilibrium will be disturbed and a temperature difference will be built up. Since the coefficient of thermal osmosis l_{qm} can be positive or negative, heat may flow in either direction.

We will now consider the general case that a temperature as well as a pressure difference exists between the two partial systems. Therefore, we may expect a superposition of the individual effects caused by the temperature and pressure differences. In some approximations we should expect a linear combination as given by

$$\begin{aligned}
\dot{Q} &= \frac{F}{d} l_q(T_2 - T_1) + \frac{F}{d} l_{qm}(p_2 - p_1) \\
\dot{M} &= \frac{F}{d} l_{mq}(T_2 - T_1) + \frac{F}{d} l_m(p_2 - p_1)
\end{aligned} \tag{7.19}$$

To give a complete survey of the problem, the entire system of equations is summarized next.

$$\begin{aligned}
\dot{E}_1 &= -\dot{E}_2 = h_1 \dot{M} + \dot{Q} \\
\dot{M}_1 &= -\dot{M}_2 = \dot{M} \\
\dot{M} &= \frac{F}{d} l_{mq}(T_2 - T_1) + \frac{F}{d} l_m(p_2 - p_1) \\
\dot{Q} &= \frac{F}{d} l_q(T_2 - T_1) + \frac{F}{d} l_{qm}(p_2 - p_1) \\
T_i &= T(E_i, M_i, V_i = \text{constant}) \\
p_i &= p(E_i, M_i, V_i = \text{constant}), \qquad i = 1, 2
\end{aligned} \tag{7.20}$$

It will be recognized immediately that this equation system is sufficient to determine the variables $\dot{M}$, $\dot{Q}$, M_1, M_2, E_1, E_2, T_1, T_2, p_1, p_2, that is we have ten equations with ten unknowns. The solution of the problem is somewhat involved. We will not attempt to actually solve the problem. All that we attempted to do was to gain a physical understanding of joint heat and mass flow.

As the final part of the problem, we will calculate the entropy change of the system. We will discover that certain relations exist among the coefficients in the material or constitutive equations (7.19). The same type of relations will appear in the more general theory to be discussed later in this chapter. Since the complete system is isolated, $d_e S = 0$, the entire entropy change is caused by internal changes which must be a positive quantity, $d_i S > 0$. We recall that the entropy is an extensive quantity. Therefore, we may add the entropies produced in the partial systems according to (7.8b) to obtain the total entropy gain. After a little algebra we obtain

$$\frac{d_i S}{dt} = \frac{dS_1}{dt} + \frac{dS_2}{dt} = \frac{1}{T_1 T_2}(T_2 - T_1)\dot{Q} + \left(s_1 - s_2 + \frac{h_2 - h_1}{T_2}\right)\dot{M} \quad (7.21)$$

To begin with, we have assumed that T_1 and T_2 do not differ too strongly. Therefore, we approximate T_2 by the mean temperature of the system T_m. Using the definition of the specific Gibbs function, the second term in (7.21) then reads

$$s_1 - s_2 + \frac{1}{T_m}(h_2 - h_1) = \frac{1}{T_m}[\mu(p_2, T_m) - \mu(p_1, T_m)] = \frac{\Delta\mu}{T_m} \quad (7.22)$$

The Taylor expansion of $\Delta\mu$, discontinuing after the linear term, gives

$$\Delta\mu = \left(\frac{\partial\mu}{\partial p}\right)_{T_m, p_m}(p_2 - p_1) = v(p_m, T_m)(p_2 - p_1) \quad (7.23)$$

Approximating the product $T_1 T_2$ by T_m^2 results in

$$\frac{d_i S}{dt} = \frac{1}{T_m^2}(T_2 - T_1)\dot{Q} + \frac{v(T_m, p_m)}{T_m}(p_2 - p_1)\dot{M} > 0 \quad (7.24)$$

This is a sum of products. Each individual product vanishes at equilibrium ($p_1 = p_2$, $T_1 = T_2$). It should be noted that in each product the terms ($\dot{Q}$, $\dot{M}$) change sign with the reversal of time. These are the thermodynamic quantities describing the transport. Their coefficients in the entropy production

$$\boxed{X_q = \frac{1}{T_m^2}(T_2 - T_1), \qquad X_m = \frac{v(p_m, T_m)}{T_m}(p_2 - p_1)} \quad (7.25)$$

are called the *conjugated thermodynamic forces* which do not change sign with the reversal of time.

To conform with meteorological convention we will divide $\dot{Q}, \dot{M}$ by the surface area F and call the resulting expressions the heat (J_F^q) and mass (J_F^m) fluxes.

$$J_F^q = \frac{\dot{Q}}{F}, \qquad J_F^m = \frac{\dot{M}}{F} \tag{7.26}$$

(Unfortunately, there is no general agreement in terminology. Some authors call the transport quantities $\dot{Q}$-fluxes and $\dot{M}$-fluxes. If these are taken per unit area, they call them flux densities). Using the abbreviations (7.25) and (7.26) we may write in place of (7.19)

$$J_F^q = L^q X_q + L^{qm} X_m, \qquad J_F^m = L^{mq} X_q + L^m X_m \tag{7.27}$$

with

$$L^q = \frac{l_q T_m^2}{d}, \qquad\qquad L^m = \frac{l_m T_m}{v(p_m, T_m)d}$$

$$L^{qm} = \frac{l_{qm} T_m}{v(p_m, T_m)d}, \qquad L^{mq} = \frac{l_{mq} T_m^2}{d} \tag{7.28}$$

Summarizing, we may state: the constitutive or *material equations of a thermodynamic system* are relations between thermodynamic (driving) forces and fluxes for the nonequilibrium state. The thermodynamic forces are invariable to changes in time while fluxes (transport quantities) change sign if time is reversed. Onsager (1931) was the first to point out these relations. The coefficients of the forces are called *phenomenological coefficients*. To guarantee that the entropy production is always positive definite, the fluxes must uniquely depend on the forces.

There exist certain relations among the phenomenological coefficients. To recognize this we substitute (7.27) into (7.24) and obtain

$$\frac{1}{F} \frac{d_i S}{dt} = L^q X_q^2 + (L^{qm} + L^{mq}) X_q X_m + L^m X_m^2 > 0 \tag{7.29}$$

This is the entropy production (with reference to the cross-section of the separating wall) which must be positive definite. Equation (7.29) can be written down concisely by using matrix notation. Defining the vector representing the forces by $\mathbf{X}$ and $\mathbf{X}^T$ by its transpose, we may write

$$\mathbf{X}^T A \mathbf{X} > 0 \quad \text{with} \quad \mathbf{X} = \begin{pmatrix} X_q \\ X_m \end{pmatrix}, \quad A = \begin{pmatrix} L^q & \dfrac{L^{qm} + L^{mq}}{2} \\ \dfrac{L^{qm} + L^{mq}}{2} & L^m \end{pmatrix}$$

$$\tag{7.30}$$

where A is a symmetric matrix. From the positivity of the entropy production follows that the matrix must be positive definite.

There exist various sufficient and necessary conditions for the real symmetric matrix to be positive definite. The conditions most useful to us are stated in the Appendix to this chapter. Due to the simplicity of A in this example, these conditions can be found very easily. Since $L^q > 0$, $L^m > 0$ we have

$$
\boxed{
\begin{array}{ll}
\text{(a)} & L^q > 0 \\[4pt]
\text{(b)} & L^m > 0 \\[4pt]
\text{(c)} & L^q L^m > \dfrac{1}{4}(L^{qm} + L^{mq})^2
\end{array}
}
\tag{7.31}
$$

Condition (7.31a) implies that in isobaric systems heat flows from the warmer to the colder system and (7.31b) implies that in isothermal systems mass will stream from the system of higher to the system of lower pressure.

Initially, empirical evidence has shown that the matrix of the system (7.27) has a very important property. It is symmetrical, i.e.

$$
L^{qm} = L^{mq} \implies l_{qm} = v_m T_m l_{mq}
\tag{7.32}
$$

This is an example of Onsager's celebrated reciprocity theorem which was derived with the help of the fluctuation theory as discussed in textbooks on statistical physics. One practical consequence of the Onsager relations is a reduction in the number of experiments to determine the phenomenological coefficients.

Before we begin with the formulation of atmospheric fluxes which depend not only on time but also on the spatial coordinates, we will study one more example to develop a deeper understanding of the theory of irreversible fluxes. With these two examples as background, we will be well prepared to understand the formalism that handles atmospheric problems.

7.2 Entropy production due to heat, mass and work exchange

As in the previous example, we treat two discontinuous systems. Each of these is homogeneous so that the thermodynamic variables depend only on time and not on spatial variables. The partial systems, characterized by the same phase, are assumed to be closed with respect to mass. However, in contrast to the previous example, they are permitted to exchange heat and work with the external surroundings, see Figure 7.2.

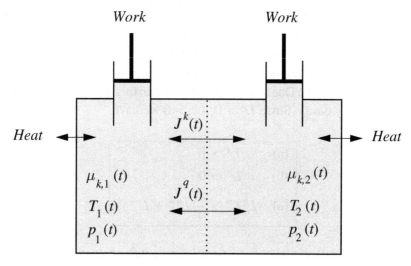

Fig. 7.2 Heat and mass exchange within a system which is closed with respect to mass.

Each subsystem ($j = 1, 2$) may also consist of n mass components M_j^k, $k = 1, 2, \ldots, n$ so that the two systems are characterized by E_j, V_j and M_j^k. Chemical reactions are excluded from the considerations. If the entropy of each subsystem is known as a function of these variables, then the intensive quantities p_j, T_j and $\mu_{k,j}$ can be found with the help of Gibbs' fundamental equation,

$$T_j dS_j = dE_j + p_j dV_j - \sum_{k=1}^{n} \mu_{k,j} dM_j^k, \qquad j = 1, 2 \qquad (7.33)$$

Complete equilibrium between the subsystems exists if

$$T_1 = T_2, \qquad p_1 = p_2, \qquad \mu_{k,1} = \mu_{k,2}, \qquad k = 1, 2, \ldots, n \qquad (7.34)$$

If only one of these equalities is violated, transport processes will take place striving to establish the equalization of the variables. As in the previous example, we require that the difference between the intensive variables $\psi = p, T, \mu_k$ is small as stated in (7.1). Again it is required that the transport between the systems is slow in comparison to equalization processes within each subsystem. If this condition is not satisfied, then all processes must be treated in terms of time and spatial variables as required in meteorological applications.

The physical laws to describe the transport between the systems are based on conservation of mass and energy.

(i) The total system is closed with respect to mass, therefore,

$$M_1^k + M_2^k = \text{constant} \implies \dot{M}_1^k = -\dot{M}_2^k = \dot{M}^k = J^k, \qquad k = 1, 2, \ldots, n \quad (7.35)$$

where J^k represents the mass flow between the partial systems.

(ii) In contrast to the first example, the internal energy may also change by the amount $d_e E_j$ caused by the interaction with the external surroundings as given by

$$\frac{d_e E_j}{dt} = \frac{đ Q_{e,j}}{dt} - p_j \frac{dV_j}{dt}, \qquad j = 1, 2 \qquad (7.36)$$

The internal interaction is formulated analogously to equation (7.8a)

$$\frac{d_i E_j}{dt} = \sum_{k=1}^{n} h_{k,j} \frac{dM_j^k}{dt} + \frac{đ Q_{im,j}}{dt}, \qquad j = 1, 2 \qquad (7.37)$$

where $h_{k,j}$ is the specific partial enthalpy. The term $đ Q_{im,j}$ is the immaterial part of $d_i E_j$. Since within the total system internal energy cannot be produced by internal processes, we have

$$\frac{d_i E_1}{dt} + \frac{d_i E_2}{dt} = 0 \qquad (7.38)$$

Denoting $đ Q_{im,j}/dt = \dot{Q}_{im,j} = J^q$ we then have

$$\frac{d_i E_1}{dt} = \sum_{k=1}^{n} h_{k,1} J^k + J^q, \qquad \frac{d_i E_2}{dt} = -\sum_{k=1}^{n} h_{k,2} J^k + \dot{Q}_{im,2} \qquad (7.39)$$

From these expressions follows immediately

$$\dot{Q}_{im,2} = -J^q + \sum_{k=1}^{n} (h_{k,2} - h_{k,1}) J^k \qquad (7.40)$$

We observe that in this case J^q and J^k do not refer to unit area of the cross-section of the dividing wall. It is a matter of choice if the transported quantities are referenced to unit area or not. Furthermore, (7.40) is a generalization of (7.12) showing that the heat fluxes are equal and opposite only if the mass flux is zero or if $h_{k,1} = h_{k,2}$ for all k.

So far no information is available about J^q and J^k. The constitutive equations for the transport processes can be obtained, however, from the entropy production within the complete system. We will show how to proceed.

The change of the entropy of the complete system is the sum of the changes of the individual parts,

$$\frac{dS}{dt} = \frac{dS_1}{dt} + \frac{dS_2}{dt} \qquad (7.41)$$

To find expressions for dS_j, $(j = 1, 2)$ from Gibbs' fundamental equation (7.33) in prognostic form, we recall that every extensive quantity can be split into external (d_e) and internal (d_i) parts so that

$$\frac{dE_j}{dt} = \frac{d_e E_j}{dt} + \frac{d_i E_j}{dt} = \frac{đ Q_{e,j}}{dt} - p_j \frac{dV_j}{dt} + \frac{d_i E_j}{dt}, \qquad j = 1, 2 \qquad (7.42)$$

Substituting (7.42) into (7.33) and using (7.41) we obtain the following expression

$$
\begin{aligned}
\frac{dS}{dt} &= \frac{d_e S}{dt} + \frac{d_i S}{dt} \\
&= \frac{1}{T_1}\frac{d\!\!\!/\, Q_{e,1}}{dt} + \frac{1}{T_2}\frac{d\!\!\!/\, Q_{e,2}}{dt} + \left(\frac{1}{T_1} - \frac{1}{T_2}\right)\frac{d_i E_1}{dt} - \sum_{k=1}^{n}\left(\frac{\mu_{k,1}}{T_1} - \frac{\mu_{k,2}}{T_2}\right)J^k
\end{aligned}
$$

(7.43)

where use has been made of the conservation equations (7.35) and (7.38). Finally, we substitute the first expression of (7.39) into (7.43) and find

(a)
$$
\frac{d_e S}{dt} = \frac{1}{T_1}\frac{d\!\!\!/\, Q_{e,1}}{dt} + \frac{1}{T_2}\frac{d\!\!\!/\, Q_{e,2}}{dt}
$$

(b)
$$
\frac{d_i S}{dt} = \left(\frac{1}{T_1} - \frac{1}{T_2}\right)\left(J^q + \sum_{k=1}^{n} h_{k,1} J^k\right) - \sum_{k=1}^{n}\left(\frac{\mu_{k,1}}{T_1} - \frac{\mu_{k,2}}{T_2}\right)J^k > 0
$$

(7.44)

Equation (7.44a) states that the external entropy influx vanishes when the individual systems do not exchange heat with the surroundings. The internal entropy production due to transport processes between the partial systems 1 and 2 vanishes identically, as indicated by (7.44b), when the heat flows J^q and the mass flows J^k are zero, which is the case when the two partial systems are isolated. If this is not the case then $d_i S/dt > 0$.

In order to obtain an approximate expression for (7.44b) we will introduce the following notation

$$
\begin{aligned}
T_1 &= T, & T_2 &= T + \Delta T \\
p_1 &= p, & p_2 &= p + \Delta p \\
\mu_{k,1} &= \mu_k, & \mu_{k,2} &= \mu_k + \delta\mu_k, & k = 1, 2, \ldots, n
\end{aligned}
$$

(7.45)

According to our previous assumption the differences in p, T and μ_k are assumed to be small quantities. Therefore, we may write

$$
\begin{aligned}
\frac{1}{T_1} - \frac{1}{T_2} &= \frac{\Delta T}{T^2}\left(1 - \frac{\Delta T}{T} + \frac{(\Delta T)^2}{T^2}\right) = \frac{\Delta T}{T^2} + \mathcal{O}(\Delta T^2) \\
\frac{\mu_{k,1}}{T_1} - \frac{\mu_{k,2}}{T_2} &= \mu_k\frac{\Delta T}{T^2} - \frac{\delta\mu_k}{T} + \mathcal{O}(\Delta T^2)
\end{aligned}
$$

(7.46)

where $\mathcal{O}(\Delta T^2)$ denotes second-order accuracy. We will now obtain suitable expressions for the chemical potentials. In general we have

$$
\mu_k = \mu_k(p, T, m^l)
$$

(7.47)

from which follows the variation

$$\delta \mu_k = \left(\frac{\partial \mu_k}{\partial T}\right)_{p,m^l} \delta T + (\delta \mu_k)_T = -s_k \delta T + (\delta \mu_k)_T$$

$$= -\frac{h_k - \mu_k}{T} \delta T + (\delta \mu_k)_T \tag{7.48}$$

$$\text{with}\quad (\delta \mu_k)_T = \left(\frac{\partial \mu_k}{\partial p}\right)_{T,m^l} \delta p + \left(\frac{\partial \mu_k}{\partial m^l}\right)_{p,T} \delta m^l$$

where $(\delta \mu_k)_T$ is that part of μ_k where the temperature is held fixed. With $\delta = d \approx \Delta$ we obtain from (7.44b) and (7.46) after some simple manipulations the expression

$$\frac{d_i S}{dt} = \frac{\Delta T}{T^2} J^q + \frac{1}{T} \sum_{k=1}^{n} (\Delta \mu_k)_T J^k \tag{7.49}$$

We introduce the symbols

$$X_q = \frac{\Delta T}{T^2}, \qquad X_k = \frac{(\Delta \mu_k)_T}{T} \tag{7.50}$$

which are the time-invariant forces and finally obtain

$$\frac{d_i S}{dt} = X_q J^q + \sum_{k=1}^{n} X_k J^k > 0 \tag{7.51}$$

Again, the entropy production is a sum of products of *thermodynamic forces* and transport quantities. The first term represents the heat exchange between the subsystems and the sum gives the exchange of mass.

According to Onsager's principle we obtain a complete set of constitutive equations if all the transport quantities (fluxes) in the entropy production equation are set up as functions of all thermodynamic forces. In this simple example we have

$$J^q = J^q(X_q, X_k), \qquad J^k = J^k(X_q, X_l), \qquad k,l = 1, 2, \ldots, n \tag{7.52}$$

Since we postulated relatively weak thermodynamic forces, see (7.1), it will be sufficient to expand (7.52) in a Taylor expansion and discontinue the expansion after the linear term. Since in equilibrium

$$J^q(0, 0) = 0, \qquad J^k(0, 0) = 0 \tag{7.53}$$

we obtain

$$J^q = \left.\frac{\partial J^q}{\partial X_q}\right|_0 X_q + \sum_{l=1}^{n} \left.\frac{\partial J^q}{\partial X_l}\right|_0 X_l$$

$$J^k = \left.\frac{\partial J^k}{\partial X_q}\right|_0 X_q + \sum_{l=1}^{n} \left.\frac{\partial J^k}{\partial X_l}\right|_0 X_l \tag{7.54}$$

Introducing abbreviations for the partial derivatives, we obtain the following expressions for the fluxes

$$J^q = L^q X_q + \sum_{k=1}^{n} L^{qk} X_k, \qquad J^k = L^{kq} X_q + \sum_{l=1}^{n} L^{kl} X_l \quad \text{with}$$

$$L^q = \frac{\partial J^q}{\partial X_q}\Big|_0, \qquad L^{qk} = \frac{\partial J^q}{\partial X_k}\Big|_0, \qquad L^{kq} = \frac{\partial J^k}{\partial X_q}\Big|_0, \qquad L^{kl} = \frac{\partial J^k}{\partial X_l}\Big|_0$$

$$(7.55)$$

The transport or phenomenological coefficients, in general, are functions of the intensive variables p, T and m^l and depend on the properties of the separating wall. As in example 1, the transport coefficients must satisfy certain inequalities. To recognize this we substitute (7.55) into (7.51) and obtain

$$\frac{d_i S}{dt} = \mathbf{X}^T A \mathbf{X} > 0 \quad \text{with} \quad \mathbf{X} = \begin{pmatrix} X_q \\ X_1 \\ \vdots \\ X_n \end{pmatrix} \qquad (7.56)$$

For the simple case that $n = 2$, the symmetrized matrix is given by

$$A = \begin{pmatrix} L^q & \dfrac{L^{q1} + L^{1q}}{2} & \dfrac{L^{q2} + L^{2q}}{2} \\ \dfrac{L^{q1} + L^{1q}}{2} & L^{11} & \dfrac{L^{12} + L^{21}}{2} \\ \dfrac{L^{q2} + L^{2q}}{2} & \dfrac{L^{12} + L^{21}}{2} & L^{22} \end{pmatrix} \qquad (7.57)$$

From the theorem listed in the Appendix to this chapter about the positive definiteness of the matrix A, we find without difficulty

$$\boxed{\begin{array}{c} L^q > 0, \qquad L^{11} > 0, \qquad L^{22} > 0 \\[2mm] L^{11}L^{22} > \left(\dfrac{L^{12} + L^{21}}{2}\right)^2, \qquad L^q L^{kk} > \left(\dfrac{L^{qk} + L^{kq}}{2}\right)^2, \qquad k = 1, 2 \end{array}}$$

$$(7.58)$$

The transport coefficients must also satisfy Onsager's reciprocity theorem stating that the coefficient matrix of the system (7.55) must be symmetrical $L^{qk} = L^{kq}, L^{kl} = L^{lk}, k, l = 1, 2$. The material equations may then be written as

$$J^q = L^q \frac{\Delta T}{T^2} + \sum_{k=1}^{n} L^{qk} \frac{(\Delta \mu_k)_T}{T}$$

$$J^k = L^{kq} \frac{\Delta T}{T^2} + \sum_{l=1}^{n} L^{kl} \frac{(\Delta \mu_l)_T}{T} \tag{7.59}$$

We will now return to our original problem and formulate the theory of irreversible fluxes to be applicable to atmospheric problems.

7.3 Meteorological applications of the theory of irreversible fluxes

7.3.1 Rearrangement of the entropy production equation

We are now ready to find expressions for the fluxes I^k, $\mathbf{J}^k$, $\mathbf{J}^q$ and $\mathbb{J}$. The starting point of the development is the entropy production equation (4.24), which will be restated for convenience

$$TQ_s = -a_{21}I^2 - a_{31}I^3 - \mathbf{J}^h \cdot \frac{\nabla T}{T} - T\mathbf{J}^n \cdot \nabla \left(\frac{\mu_n}{T} \right) + \mathbb{J} \cdot\cdot \nabla \mathbf{v} \geq 0 \tag{7.60}$$

It should be carefully noted that in this equation (in contrast to the examples) entropy production now refers to unit volume so that TQ_s in the m–kg–s system is given in units of (Joule s^{-1} m^{-3}). It will also be noted that this equation contains the heat flux while the sensible heat flux is required in the formulation of the first law of thermodynamics (3.52). According to (3.51) these two fluxes are related by

$$\mathbf{J}_s^h = \mathbf{J}^h - \mathbf{J}^n h_n \tag{7.61}$$

We will now eliminate $\mathbf{J}^h$ in (7.60) and replace it by $\mathbf{J}_s^h$. Before proceeding, we recall the relationship (7.48) and replace the variational symbol by the gradient operator so that

$$\nabla \mu_k = -(h_k + \mu_k) \frac{\nabla T}{T} + (\nabla \mu_k)_T \tag{7.62}$$

Since

$$T\nabla \left(\frac{\mu_k}{T} \right) = \nabla \mu_k - \mu_k \frac{\nabla T}{T} \tag{7.63}$$

we may rewrite (7.62) and obtain

$$T\nabla \left(\frac{\mu_k}{T} \right) = -h_k \frac{\nabla T}{T} + (\nabla \mu_k)_T \tag{7.64}$$

Now we substitute (7.64) into (7.60) and use (7.61). This results in

$$TQ_s = -a_{21}I^2 - a_{31}I^3 - \mathbf{J}_s^h \cdot \frac{\nabla T}{T} - \mathbf{J}^n \cdot (\nabla \mu_n)_T + \mathbb{J} \cdot\cdot \nabla \mathbf{v} \geq 0 \qquad (7.65)$$

As in the previous chapters, in (7.65) we eliminate $\mathbf{J}^0$ by means of $\mathbf{J}^0 = -(\mathbf{J}^1 + \mathbf{J}^2 + \mathbf{J}^3)$. This leads to the following expression for the entropy production

$$\begin{aligned} TQ_s = &-a_{21}I^2 - a_{31}I^3 - \mathbf{J}_s^h \cdot \frac{\nabla T}{T} - \mathbf{J}^1 \cdot \nabla(\mu_1 - \mu_0)_T \\ &- \mathbf{J}^2 \cdot \nabla(\mu_2 - \mu_0)_T - \mathbf{J}^3 \cdot \nabla(\mu_3 - \mu_0)_T + \mathbb{J} \cdot\cdot \nabla \mathbf{v} \geq 0 \end{aligned} \qquad (7.66)$$

Next we rearrange the dissipation term in a two-step operation.

(1) We decompose the symmetric dyadic $\mathbb{J}$

$$\mathbb{J} = J\mathbb{E} + \mathbb{J}^o \qquad (7.67)$$

with $J\mathbb{E}$ the isotropic symmetric part and $\mathbb{J}^o$ the anisotropic symmetric part. To find J, we carry out the indicated operation

$$\mathbb{J} \cdot\cdot \mathbb{E} = J\mathbb{E} \cdot\cdot \mathbb{E} + \mathbb{J}^o \cdot\cdot \mathbb{E} = 3J, \implies J = \frac{\mathbb{J} \cdot\cdot \mathbb{E}}{3} \qquad (7.68)$$

(2) We decompose the local velocity dyadic into its symmetric part $\mathbb{V}^s$ and its anti-symmetric part $\mathbb{V}^a$.

$$\nabla \mathbf{v} = \mathbb{V}^s + \mathbb{V}^a \quad \text{with}$$
$$\mathbb{V}^s = \frac{1}{2}(\nabla \mathbf{v} + \overset{\frown}{\mathbf{v}\nabla}), \qquad \mathbb{V}^a = \frac{1}{2}(\nabla \mathbf{v} - \overset{\frown}{\mathbf{v}\nabla}) \qquad (7.69)$$

Next we decompose the symmetric part of the local velocity dyadic in analogy to the decomposition of $\mathbb{J}$.

$$\mathbb{V}^s = V\mathbb{E} + \mathbb{V}^o, \qquad V = \frac{1}{3}\nabla \cdot \mathbf{v} \qquad (7.70)$$

Therefore, we obtain

$$\mathbb{V}^o = \mathbb{V}^s - \frac{\nabla \cdot \mathbf{v}}{3}\mathbb{E} = \frac{1}{2}\left(\nabla \mathbf{v} + \overset{\frown}{\mathbf{v}\nabla} - \frac{2}{3}\nabla \cdot \mathbf{v}\mathbb{E}\right) \qquad (7.71)$$

Observing that the double scalar products of the following dyadics vanish

$$\mathbb{E} \cdot\cdot \mathbb{V}^o = 0, \qquad \mathbb{E} \cdot\cdot \mathbb{V}^a = 0, \qquad \mathbb{J}^o \cdot\cdot \mathbb{E} = 0, \qquad \mathbb{J}^o \cdot\cdot \mathbb{V}^a = 0 \qquad (7.72)$$

Table 7.1. *Fluxes and thermodynamic forces*

Flux	Thermodynamic force
Scalars	
$J^{(1)} = J$	$a_{(1)} = \nabla \cdot \mathbf{v}$
$J^{(2)} = I^2$	$a_{(2)} = -a_{21}$
$J^{(3)} = I^3$	$a_{(3)} = -a_{31}$
Vectors	
$\mathbf{J}^{(0)} = \mathbf{J}_s^h$	$\mathbf{a}_{(0)} = -\nabla T/T$
$\mathbf{J}^{(1)} = \mathbf{J}^1$	$\mathbf{a}_{(1)} = -\nabla(\mu_1 - \mu_0)_T$
$\mathbf{J}^{(2)} = \mathbf{J}^2$	$\mathbf{a}_{(2)} = -\nabla(\mu_2 - \mu_0)_T$
$\mathbf{J}^{(3)} = \mathbf{J}^3$	$\mathbf{a}_{(3)} = -\nabla(\mu_3 - \mu_0)_T$
Tensor	
$\mathbb{J}^0$	$\mathbb{A}_0 = \frac{1}{2}(\nabla \mathbf{v} + \widehat{\mathbf{v}\nabla} - \frac{2}{3}\nabla \cdot \mathbf{v}\mathbb{E}) = \mathbb{V}^0$

we finally obtain the result

$$\mathbb{J} \cdot\cdot \nabla \mathbf{v} = (J\mathbb{E} + \mathbb{J}^o) \cdot\cdot \left(\frac{1}{3}\nabla \cdot \mathbf{v}\mathbb{E} + \mathbb{V}^o + \mathbb{V}^a \right)$$

$$= J\nabla \cdot \mathbf{v} + \mathbb{J}^o \cdot\cdot \mathbb{V}^o \qquad (7.73)$$

Substituting this expression into (7.66) yields the final form of the entropy production equation

$$TQ_s = J\nabla \cdot \mathbf{v} - \sum_{k=2}^{3} I^k a_{k1} - \mathbf{J}_s^h \cdot \frac{\nabla T}{T} - \sum_{k=1}^{3} \mathbf{J}^k \cdot \nabla(\mu_k - \mu_0)_T + \mathbb{J}^o \cdot\cdot \mathbb{V}^o \geq 0$$

$$(7.74)$$

As in the previous examples, the entropy production presents itself as a sum of products. One factor in each product is one of the as yet undetermined fluxes while the second factor represents the conjugated *thermodynamic force*.

To get a unified notation for fluxes and forces, we will introduce the symbols as listed in Table 7.1. We will decompose the general expression TQ_s into a part consisting of scalars only $(TQ_s)_S$, a part consisting of vectors only $(TQ_s)_V$ and a tensorial part $(TQ_s)_T$ with

$$(TQ_s)_S = \sum_{k=1}^{3} J^{(k)} a_{(k)}, \quad (TQ_s)_V = \sum_{k=0}^{3} \mathbf{J}^{(k)} \cdot \mathbf{a}_{(k)}, \quad (TQ_s)_T = \mathbb{J}^o \cdot\cdot \mathbb{A}_0 \quad (7.75)$$

so that

$$TQ_s = (TQ_s)_S + (TQ_s)_V + (TQ_s)_T \geq 0 \qquad (7.76)$$

7.3.2 Relation between fluxes and thermodynamic forces

Since we are interested in the irreversible fluxes we need to consider only $Q_s > 0$. As the two examples indicate, the positive definiteness of the entropy production can be guaranteed only by unique relations between fluxes and the conjugated thermodynamic forces. Otherwise combinations of fluxes and forces might occur which could be in contradiction to the second law of thermodynamics yielding $Q_s < 0$. In this section we will present the *theory of irreversible fluxes* as far as needed for our work. We will soon find out that the information gained from the two examples will have to be extended to handle atmospheric problems.

The theory of irreversible fluxes is based on two fundamental requirements.

7.3.2.1 Linearity

For not too strong deviations from the thermodynamic equilibrium, i.e. Q_s must not be too large, the fluxes and their conjugated thermodynamic forces must be related in a linear fashion. In general, each flux depends on every thermodynamic force.

7.3.2.2 Principle of Curie

A basic assumption is that the fluid medium is locally isotropic, i.e. its properties are assumed to be independent of direction. The atmosphere can certainly be regarded as such a medium. Thus, the phenomenological equations must be invariant to rotation of the reference system. According to P. Curie then the fluxes depend linearly only on such forces which under rotation in space show the same transformation properties. Three important properties follow.

(i) Scalar fluxes depend only on scalar thermodynamic forces.
(ii) Vectorial fluxes depend only on vectorial thermodynamic forces.
(iii) Tensorial fluxes depend only on tensorial thermodynamic forces.

If the medium, for example, were to be a nonisotropic polarizable fluid under the influence of an external electric field, no such simplifications would occur. Therefore, we can write

$$\boxed{(TQ_s)_\text{S} > 0, \qquad (TQ_s)_\text{V} > 0, \qquad (TQ_s)_\text{T} > 0} \tag{7.77}$$

and for the flux relations

$$J^{(k)} = \sum_{i=1}^{3} l^{ki} a_{(i)}, \qquad \mathbf{J}^{(k)} = \sum_{i=0}^{3} L^{ki} \mathbf{a}_{(i)}, \qquad \mathbb{J}^o = 2\widetilde{\mu} \mathbb{A}_0 \tag{7.78}$$

The factor 2 in the last expression is irrelevant but is convenient.

As was shown in the previous two examples, the phenomenological coefficients of (7.78) cannot be specified independently, but they must obey certain inequality relations. To determine these, we substitute (7.78) into the corresponding equations (7.75) and obtain the positive definite forms of the entropy production.

Substitution of the first expression of (7.78) into (7.75) gives

$$(TQ_s)_S = \mathbf{X}_S^T A_S \mathbf{X}_S > 0 \quad \text{with} \quad \mathbf{X}_S = \begin{pmatrix} a_{(1)} \\ a_{(2)} \\ a_{(3)} \end{pmatrix} \tag{7.79}$$

The symmetrized matrix A_S is written out for convenience to more easily recognize the relations among the coefficients

$$A_S = \begin{pmatrix} l^{11} & \dfrac{l^{12} + l^{21}}{2} & \dfrac{l^{13} + l^{31}}{2} \\ \dfrac{l^{12} + l^{21}}{2} & l^{22} & \dfrac{l^{23} + l^{32}}{2} \\ \dfrac{l^{13} + l^{31}}{2} & \dfrac{l^{23} + l^{32}}{2} & l^{33} \end{pmatrix} \tag{7.80}$$

Since the matrix is positive definite every principal minor must be positive definite, i.e.

$$l^{11} > 0, \qquad l^{22} > 0, \qquad l^{33} > 0$$

$$l^{11}l^{22} > \frac{(l^{12} + l^{21})^2}{4}, \qquad l^{11}l^{33} > \frac{(l^{13} + l^{31})^2}{4} \tag{7.81}$$

$$l^{22}l^{33} > \frac{(l^{23} + l^{32})^2}{4}, \qquad |(A_S)| > 0$$

Substituting the vectorial fluxes of (7.78) into (7.75) yields

$$(TQ_s)_V = \mathbf{X}_V^T A_V \mathbf{X}_V \quad \text{with}$$

$$\mathbf{X} = \begin{pmatrix} a_{(0)} \\ a_{(1)} \\ a_{(2)} \\ a_{(3)} \end{pmatrix}, \qquad A_V = \begin{pmatrix} L^{00} & L^{01} & L^{02} & L^{03} \\ L^{10} & L^{11} & L^{12} & L^{13} \\ L^{20} & L^{21} & L^{22} & L^{23} \\ L^{30} & L^{31} & L^{32} & L^{33} \end{pmatrix} \tag{7.82}$$

Since this matrix is also positive definite every principle minor must be positive definite

$$L^{kk} > 0, \quad k = 0, \dots, 3, \qquad L^{ii}L^{kk} > \frac{(L^{ik} + L^{ki})^2}{4}, \qquad i, k = 0, \dots, 3, \quad i \neq k$$

$$\begin{vmatrix} L^{00} & L^{01} & L^{02} \\ L^{10} & L^{11} & L^{12} \\ L^{20} & L^{21} & L^{22} \end{vmatrix} > 0, \qquad \begin{vmatrix} L^{11} & L^{12} & L^{13} \\ L^{21} & L^{22} & L^{23} \\ L^{31} & L^{32} & L^{33} \end{vmatrix} > 0, \qquad \begin{vmatrix} A_V \end{vmatrix} > 0$$

$$\tag{7.83}$$

From the two last expressions of (7.75) and (7.78) we see that the coefficient $\tilde{\mu}$ must be positive.

7.3.3 Onsager–Casimir reciprocity relations

In some cases the phenomenological coefficients can be calculated from statistical thermodynamics. In general, these coefficients will be determined experimentally. Let v^{ik} represent both types of coefficients l^{ik} and L^{ik}. The v^{ik} possess certain symmetry properties which can be obtained from the theoretical investigations of Onsager and Casimir. To actually derive the symmetry properties is quite involved. Therefore, we will state the results only.

All thermodynamic forces (or fluxes) can be associated with a number describing the behavior of a particular force with the reversal of time. This number is called the parity of the force which is $+1$ if the force is invariant and -1 if it is not invariant to time reversal. The general law can be written as

$$v^{ik} = \epsilon_i \epsilon_k v^{ki}, \qquad \epsilon_i = \pm 1, \qquad \epsilon_k = \pm 1 \tag{7.84}$$

where $\epsilon_i, \epsilon_k = -1$ if forces change sign for $t \to -t$ and $\epsilon_i, \epsilon_k = +1$ if forces do not change sign for $t \to -t$. For rotating systems this equation must be modified.

The principle of Curie made it possible to reduce the number of the independent coefficients in the linear constitutive equations. The Onsager–Casimir reciprocal relations lead to an additional reduction in the number of the coefficients in the linear laws.

We will now apply equation (7.84) to the phenomenological coefficients. Substituting the fluxes $J^{(k)}$ and $\mathbf{J}^{(k)}$ as defined by (7.78) into (7.75) we obtain

$$
\begin{aligned}
(TQ_s)_{\mathrm{S}} &= \sum_{k=1}^{3} \sum_{i=1}^{3} l^{ki} a_{(i)} a_{(k)} > 0 \\
(TQ_s)_{\mathrm{V}} &= \sum_{k=0}^{3} \sum_{i=0}^{3} L^{ki} \mathbf{a}_{(i)} \cdot \mathbf{a}_{(k)} > 0
\end{aligned}
\tag{7.85}
$$

The force $a_{(1)} = \nabla \cdot \mathbf{v}$ is an odd force since with $t \to -t$ one obtains $\mathbf{v} \to -\mathbf{v}$ and $\nabla \cdot \mathbf{v} \to -\nabla \cdot \mathbf{v}$. The remaining forces occurring in (7.85) are even. We will now investigate the symmetry properties of each l^{ki} and L^{ki} in (7.85).

7.3.3.1 Symmetry relations of the scalar fluxes

Only the thermodynamic force $a_{(1)}$ is of parity $\epsilon_1 = -1$ while $\epsilon_2 = \epsilon_3 = +1$. Thus, from (7.85) we recognize that

Table 7.2. *Reciprocity relations of Onsager and Casimir*

Casimir relations	$l^{12} = -l^{21},\ l^{13} = -l^{31}$	
Onsager relations	$l^{23} = l^{32}$	$L^{ik} = L^{ki},\ i, k = 0, 1, 2, 3$

$$\boxed{l^{12} = -l^{21}, \qquad l^{13} = -l^{31}, \qquad l^{23} = l^{32}} \tag{7.86}$$

7.3.3.2 Symmetry relations of the vectorial fluxes

All thermodynamic forces $\mathbf{a}_{(i)}$ are invariant to time reversal $t \rightarrow -t$, i.e. all ϵ_i are $+1$, so that

$$\boxed{L^{ik} = L^{ki}, \qquad i, k = 0, \ldots, 3} \tag{7.87}$$

The phenomenological coefficients changing sign with $t \rightarrow -t$ are attributed to Casimir while the remaining coefficients bear Onsager's name. We summarize the results in Table 7.2.

7.3.4 The flux relations

We will now apply the reciprocity relations to obtain the final form of the flux equations. Instead of equation (7.78), we obtain with the help of Table 7.1 the following set of equations.

$$
\begin{aligned}
J &= l^{11} \nabla \cdot \mathbf{v} - l^{12} a_{21} - l^{13} a_{31} \\
I^2 &= -l^{12} \nabla \cdot \mathbf{v} - l^{22} a_{21} - l^{23} a_{31} \\
I^3 &= -l^{13} \nabla \cdot \mathbf{v} - l^{23} a_{21} - l^{33} a_{31} \\
\mathbf{J}_s^h &= -\frac{L^{00}}{T} \nabla T - L^{01} \nabla(\mu_1 - \mu_0)_T - L^{02} \nabla(\mu_2 - \mu_0)_T - L^{03} \nabla(\mu_3 - \mu_0)_T \\
\mathbf{J}^1 &= -\frac{L^{01}}{T} \nabla T - L^{11} \nabla(\mu_1 - \mu_0)_T - L^{12} \nabla(\mu_2 - \mu_0)_T - L^{13} \nabla(\mu_3 - \mu_0)_T \\
\mathbf{J}^2 &= -\frac{L^{02}}{T} \nabla T - L^{12} \nabla(\mu_1 - \mu_0)_T - L^{22} \nabla(\mu_2 - \mu_0)_T - L^{23} \nabla(\mu_3 - \mu_0)_T \\
\mathbf{J}^3 &= -\frac{L^{03}}{T} \nabla T - L^{13} \nabla(\mu_1 - \mu_0)_T - L^{23} \nabla(\mu_2 - \mu_0)_T - L^{33} \nabla(\mu_3 - \mu_0)_T \\
\mathbb{J}^o &= \tilde{\mu} \left(\nabla \mathbf{v} + \overset{\frown}{\mathbf{v}\nabla} - \frac{2}{3} \nabla \cdot \mathbf{v} \mathbb{E} \right)
\end{aligned}
$$

$$\tag{7.88}$$

We recognize that the application of the reciprocity relations has reduced the number of the phenomenological coefficients in the scalar fluxes from 9 to 6 and in the vectorial fluxes from 16 to 10. In general the phenomenological coefficients are functions of the intensive state variables p, T and m^k, but often we simply treat them as constants.

7.3.5 Simplified flux relations

For many practical purposes it may be sufficient to simplify the flux equations by ignoring some superpositions. As an example we consider the situation of $l^{12} = l^{13} = 0$ in (7.88). This yields a decoupling of viscous processes and phase changes. The first equation of (7.88) then reduces to

$$J = l^{11}\nabla \cdot \mathbf{v} \implies \mathbb{J} = J\mathbb{E} + \mathbb{J}^o = \widetilde{\mu}(\nabla\mathbf{v} + \widehat{\mathbf{v}\nabla}) - \left(\frac{2}{3}\widetilde{\mu} - l^{11}\right)\nabla \cdot \mathbf{v}\mathbb{E} \quad (7.89)$$

The coefficients l^{11} and $\widetilde{\mu}$ are called the coefficients of volume and shear viscosity, respectively. Often l^{11} is ignored altogether in atmospheric systems. In general, however, the dyadic or the *stress tensor* as it is often called, is also dependent on the phase changes. For the second and third expression of (7.88) we obtain with $l^{11} \approx 0$

$$I^2 = -l^{22}a_{21} - l^{23}a_{31}, \qquad I^3 = -l^{23}a_{21} - l^{33}a_{31} \quad (7.90)$$

These equations, usually in some disguised form, are a part of any cloud model.

7.4 Vectorial fluxes in moist air

In the final two sections of this chapter we will consider the processes of baro-diffusion and heat conduction in the moist but cloudless atmosphere.

For cloudless but moist air, we have to deal with the following pair of equations

$$\mathbf{J}_s^h = -\frac{L^{00}}{T}\nabla T - L^{01}\nabla(\mu_1 - \mu_0)_T$$
$$\mathbf{J}^1 = -\frac{L^{01}}{T}\nabla T - L^{11}\nabla(\mu_1 - \mu_0)_T \quad (7.91)$$

To bring (7.91) into a suitable form we make use of (7.48) and recall the identity (4.12). Thus, we obtain

$$\nabla(\mu_1 - \mu_0)_T = \frac{\partial}{\partial p}(\mu_1 - \mu_0)_{T,m^1}\nabla p + \frac{\partial}{\partial m^1}(\mu_1 - \mu_0)_{p,T}\nabla m^1$$
$$= (v_1 - v_0)\nabla p + \frac{\partial}{\partial m^1}(\mu_1 - \mu_0)_{p,T}\nabla m^1 \quad (7.92)$$

From the Gibbs–Duhem equation (6.60) applied to a system at constant pressure and temperature, as required by the second expression of (7.92), we find for $m^1 \neq 1$

$$m^0 \left(\frac{\partial \mu_0}{\partial m^1}\right)_{p,T} + m^1 \left(\frac{\partial \mu_1}{\partial m^1}\right)_{p,T} = 0 \tag{7.93}$$

or

$$\frac{\partial}{\partial m^1}(\mu_1 - \mu_0)_{p,T} = \frac{1}{1 - m^1}\left(\frac{\partial \mu_1}{\partial m^1}\right)_{p,T} \tag{7.94}$$

This equation substituted into (7.92) gives

$$\nabla(\mu_1 - \mu_0)_T = (v_1 - v_0)\nabla p + \frac{\nabla m^1}{1 - m^1}\left(\frac{\partial \mu_1}{\partial m^1}\right)_{p,T} \tag{7.95}$$

With the use of this equation the vectorial equations (7.91) can be written in the final form as

$$\mathbf{J}_s^h = -\frac{L^{00}}{T}\nabla T - L^{01}(v_1 - v_0)\nabla p - L^{01}\frac{\nabla m^1}{1 - m^1}\left(\frac{\partial \mu_1}{\partial m^1}\right)_{p,T}$$

$$\mathbf{J}^1 = -\frac{L^{01}}{T}\nabla T - L^{11}(v_1 - v_0)\nabla p - L^{11}\frac{\nabla m^1}{1 - m^1}\left(\frac{\partial \mu_1}{\partial m^1}\right)_{p,T} \tag{7.96}$$

7.4.1 Baro-diffusion

We consider an isothermal mixture of dry air and water vapor. According to (7.96) the diffusion of water vapor is then given by

$$\mathbf{J}^1 = -\mathbf{J}^0 = -L^{11}(v_1 - v_0)\nabla p - L^{11}\frac{\nabla m^1}{1 - m^1}\left(\frac{\partial \mu_1}{\partial m^1}\right)_{p,T} \tag{7.97}$$

This type of diffusion is called baro-diffusion since not only the concentration gradient of the water vapor determines the diffusion process, but the diffusion is also driven by the pressure gradient. Before we study the complete equation, it will be instructive to first consider the case of a nonexisting pressure gradient.

7.4.1.1 Water vapor diffusion due to a concentration gradient

We set $\nabla p = 0$ and study the remaining part of (7.97). The coefficient multiplying the concentration gradient of water vapor is known as the *diffusion coefficient* which will be denoted here by the symbol D^{11}. The task ahead is to relate the diffusion coefficient to the phenomenological coefficient L^{11}. In order to proceed we need to have a suitable expression for μ_1, the chemical potential of water vapor. At the present time such an expression is not at our disposal, but in the next chapter it will be derived independently of any diffusion considerations. This expression is given by

$$\mu_1 = \overset{+}{\mu}_1(T) + R_1 T \ln p + R_1 T \ln n^1 \qquad (7.98)$$

The term $\overset{+}{\mu}_1(T)$ combines several terms which are either constants or depend on temperature only. The quantities R_1 and n^1 are the gas constant and the molar fraction of water vapor. Since $N^1 = M^1/\breve{m}_1$, where $\breve{m}_1$ is the molecular weight of water vapor, we readily obtain

$$n^1 = \frac{N^1}{N^0 + N^1} = \frac{m^1/\breve{m}_1}{m^0/\breve{m}_0 + m^1/\breve{m}_1} \qquad (7.99)$$

where we have divided the numerator and denominator by the total mass $M^0 + M^1 = M$. Taking the logarithm of (7.99) and substituting the result into (7.98), we find without difficulty

$$\left(\frac{\partial \mu_1}{\partial m^1}\right)_{p,T} = \frac{R^* T}{m^1 \breve{m}_1}\left[1 - n^1\left(\frac{\breve{m}_0 - \breve{m}_1}{\breve{m}_0}\right)\right] \quad \text{with} \quad R^* = R_1 \breve{m}_1 \qquad (7.100)$$

With $1 - n^1 + n^1\breve{m}_1/\breve{m}_0 = n^1\breve{m}_1/(m^1\breve{m}_0)$ we obtain the expression

$$\frac{1}{1 - m^1}\left(\frac{\partial \mu_1}{\partial m^1}\right)_{p,T} = \frac{R^* T n^1}{m^0 \breve{m}_0 (m^1)^2} = d \qquad (7.101)$$

where the symbol d has been introduced for brevity. The diffusion expression (7.97), in the absence of the pressure gradient, can then be written as

$$\mathbf{J}^1 = -D^{11}\nabla m^1 \quad \text{with} \quad D^{11} = L^{11}d \qquad (7.102)$$

If the diffusion flux is expressed in units of $\text{kg m}^{-2}\,\text{s}^{-1}$ then D^{11} has units $\text{kg m}^{-1}\,\text{s}^{-1}$.

The diffusion coefficient D^{11} can now be used to obtain the coefficient L^{11}. D^{11} can be calculated from gas kinetic considerations or it is found from measurements. For a binary mixture of ideal gases Chapman and Cowling (1939) give an approximate formula

$$D^{11} = \sqrt{\frac{R_0 + R_1}{2\pi}}\frac{3k_B T^{3/2}\rho}{8a_{10}^2 p} > 0 \quad \text{with} \quad a_{10} = \frac{a_0 + a_1}{2} \qquad (7.103)$$

where a_0, a_1 are the diameters of the two types of molecules, ρ is the density of the gas and k_B is the Boltzmann constant. The water vapor diffusion coefficient in this approximation is a function of pressure and temperature.

7.4.1.2 Water vapor diffusion as modified by the pressure gradient

This case is known as baro-diffusion and we must write

$$\mathbf{J}^1 = -L^{11}(v_1 - v_0)\nabla p - D^{11}\nabla m^1 \qquad (7.104)$$

We will now relate the coefficient multiplying the pressure gradient to the so-called *baro-diffusion coefficient* D_p^{11} which is defined by

$$D_p^{11} = pL^{11}(v_1 - v_0) = \frac{p}{d}D^{11}(v_1 - v_0) \qquad (7.105)$$

The factor p was introduced so that D_p^{11} can be expressed in the same units as D^{11}. In case of a vanishing concentration gradient the coefficient D_p^{11} determines the quantity of water vapor transported due to the existence of the pressure gradient.

The quantity D_p^{11} can be calculated without difficulty. By utilizing (7.101) and

$$v_1 - v_0 = \frac{(R_1 - R_0)T}{p} \qquad (7.106)$$

we obtain immediately

$$D_p^{11} = D^{11}\frac{(R_1 - R_0)m^0(m^1)^2}{R_0 n^1} \qquad (7.107)$$

It is useful to introduce the so-called *baro-diffusion ratio* k_p as defined by

$$k_p = \frac{D_p^{11}}{D^{11}} = \frac{pL^{11}(v_1 - v_0)}{D^{11}} \qquad (7.108)$$

With this definition equation (7.104) can be written as

$$\mathbf{J}^1 = -D^{11}\left(\nabla m^1 + \frac{k_p}{p}\nabla p\right) \qquad (7.109)$$

In this subsection we followed a paper by Herbert (1973) where further details may be found. The same paper also gives a fundamental treatment of the diffusion of water droplets which are suspended in the air. For turbulent motion some simplified flux equations are discussed in *DA*.

7.4.2 Heat conduction

7.4.2.1 The heat conduction equation

In the final section of this chapter we will consider the situation of simple heat conduction by ignoring the superposition in the first equation of (7.91). What remains is

$$\mathbf{J}_s^h = -\frac{L^{00}}{T}\nabla T = -l\nabla T \quad \text{with} \quad l = \frac{L^{00}}{T} \tag{7.110}$$

where l is the *coefficient of heat conduction*. This equation is known as *Fourier's law*. If the conducting medium is at rest, the first law of thermodynamics can be written as

$$c_v \rho \frac{\partial T}{\partial t} = -\nabla \cdot \mathbf{J}_s^h \tag{7.111}$$

Denoting the *thermal diffusivity* by the lumped expression $\lambda = l/\rho c_v$ we obtain the *heat conduction equation*

$$\frac{\partial T}{\partial t} = \lambda \nabla^2 T \tag{7.112}$$

where we have assumed a constant value of l. For this simple linear partial differential equation of the parabolic type (first order time derivative) it is possible to obtain an analytical solution. For the more complicated atmospheric problems (involving turbulence and stability) we get a more complicated equation, which must be solved numerically.

In many situations it is expedient to define a new origin of temperature, for example by $T' = T - \overline{T}$ where $\overline{T}$ is a constant. The differential equation does not change by this transformation and the form of the initial and boundary conditions remains the same.

7.4.2.2 Temperature wave within the soil

In most atmospheric problems the upper part of the ground and the atmospheric boundary layer are considered to be a coupled system. Then we have to solve two equations of the type (7.112), one for the ground and one for the atmosphere. At the surface of the earth the two solutions must coincide. This is done by specifying a certain interface condition which contains the atmospheric and the soil heat fluxes and the energy input due to solar and atmospheric radiation. This type of problem is treated in textbooks on dynamical meteorology. Presently, we will find a solution to (7.112) by specifying the temperature wave at the earth's surface assuming that heat is conducted along the z-axis only which is directed into the ground assumed to extend to infinity.

The problem to be solved is the boundary value problem

$$\frac{\partial T'}{\partial t} = \lambda \frac{\partial^2 T'}{\partial z^2} \quad \text{with} \quad T'(z = 0, t) = T_0 \cos \omega t, \qquad \lambda = \text{constant} \tag{7.113}$$

where ω is the circular frequency ($\tau = 2\pi/\omega$ is the period of the temperature wave) and T_0 the amplitude. We solve the problem by separating the variables

$$T'(z, t) = T_1(z)T_2(t) \tag{7.114}$$

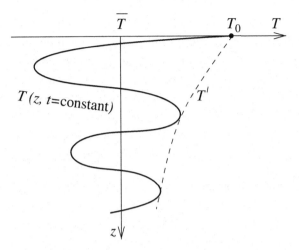

Fig. 7.3 Temperature wave within the soil.

Substitution of this equation into (7.113) gives

$$\frac{1}{\lambda T_2}\frac{dT_2}{dt} = \frac{1}{T_1}\frac{d^2 T_1}{dz^2} = c_s^2 \tag{7.115}$$

where c_s^2 is the separation constant. The partial differential equation has now been split into two ordinary differential equations. The two parts of the solution are given by

$$T_2 = c_1 \exp\left(c_s^2 \lambda t\right), \qquad T_1 = c_2 \exp(c_s z) + c_3 \exp(-c_s z) \tag{7.116}$$

and the complete solution is given by

$$T'(z, t) = \bar{c}_1 \exp\left(c_s^2 \lambda t + c_s z\right) + \bar{c}_2 \exp\left(c_s^2 \lambda t - c_s z\right) \tag{7.117}$$

The constants c_s^2, $\bar{c}_1 = c_1 c_2$ and $\bar{c}_2 = c_1 c_3$ may be complex. To satisfy the boundary condition at $z = 0$, we identify $c_s^2 = i\omega/\lambda$ and take only the real part of the solution. This gives with $\sqrt{i} = (1 + i)/\sqrt{2}$

$$\begin{aligned}\Re\left[T'(z, t)\right] = \Re&\left[\bar{c}_1 \exp\left(i\omega t + i\sqrt{\omega/2\lambda}\, z\right) \exp\left(\sqrt{\omega/2\lambda}\, z\right)\right] \\ + \Re&\left[\bar{c}_2 \exp\left(i\omega t - i\sqrt{\omega/2\lambda}\, z\right) \exp\left(-\sqrt{\omega/2\lambda}\, z\right)\right]\end{aligned} \tag{7.118}$$

To prevent the solution to approach infinity for $z \to \infty$ we set $\bar{c}_1 = 0$. The boundary condition at $z = 0$ is satisfied if we set $\bar{c}_2 = T_0$. The temperature distribution T' is then given by

$$T'(z, t) = T_0 \exp\left(-\sqrt{\omega/2\lambda}\, z\right) \cos\left(\omega t - \sqrt{\omega/2\lambda}\, z\right) \tag{7.119}$$

and the final solution is

$$T(z, t) = \overline{T} + T_0 \exp\left(-\sqrt{\omega/2\lambda}\, z\right) \cos\left(\omega t - \sqrt{\omega/2\lambda}\, z\right) \tag{7.120}$$

The temperature distribution is displayed in Figure 7.3.

7.4.2.3 Discussion of the solution

The solution indicates that the amplitude vanishes at infinity. For all practical purposes, however, the amplitude already vanishes a few meters below the earth's surface. A useful measure for the penetration of the temperature wave into the ground is the so-called *penetration depth* z_p at which the maximum amplitude has decreased to $e^{-1}T_0$. Therefore,

$$T_0 \exp(-1) = T_0 \exp\left(-\sqrt{\omega/2\lambda}\, z_p\right) \quad \text{or} \quad z_p = \sqrt{\frac{2\lambda}{\omega}} = \sqrt{\frac{\lambda\tau}{\pi}} \qquad (7.121)$$

Let the diurnal and annual period be denoted by τ_d and τ_a, respectively. Then

$$\frac{(z_p)_a}{(z_p)_d} = \sqrt{\frac{\tau_a}{\tau_d}} = \sqrt{\frac{365}{1}} \approx 19 \qquad (7.122)$$

Measurements show that on the average the diurnal temperature wave is still detectable at a depth at 1 m, the annual temperature wave penetrates to a depth of about 17 m.

7.4.2.4 Wavelength and phase speed

Let the wavelength be given by the distance $L = z_2 - z_1 > 0$. The wavelength is the distance between two points of equal phase at $t =$ constant such as the distance between two temperature maxima. Since the arguments of the cosine must differ by 2π, we have

$$2\pi = \left(\omega t - \sqrt{\frac{\omega}{2\lambda}}z_1\right) - \left(\omega t - \sqrt{\frac{\omega}{2\lambda}}z_2\right) = (z_2 - z_1)\sqrt{\frac{\omega}{2\lambda}}$$

$$L = 2\pi\sqrt{\frac{2\lambda}{\omega}} = 2\sqrt{\pi\lambda\tau} \qquad (7.123)$$

Therefore, the wavelength is proportional to the square root of the period.

The speed of a point of constant phase, such as the maximum of the temperature wave, is known as the *phase speed c*. For a constant phase the argument of the cosine is

$$\omega t - \sqrt{\frac{\omega}{2\lambda}}z = \text{constant} \qquad (7.124)$$

Therefore,

$$c = \frac{dz}{dt} = \sqrt{2\omega\lambda} = 2\sqrt{\frac{\pi\lambda}{\tau}} \qquad (7.125)$$

The speed of the wave in z-direction increases with decreasing period.

Detailed analytic solutions for various types of heat conduction problems may be found in Carslaw and Jaeger (1959). More recently, Zdunkowski and Kandlbinder

(1997) presented a fairly complete solution to a heat conduction problem where the atmosphere and the soil are treated as a coupled system.

7.5 Appendix

7.5.1 Definition 1

If A is a square matrix, any square submatrix of A whose principal diagonal is a part of the principal diagonal of A is called a principal submatrix of A. The determinant of any principal submatrix of a square matrix A is called a principal minor of A.

7.5.2 Definition 2

Let **X** be represented by the vector

$$\mathbf{X} = \begin{pmatrix} X_1 \\ \vdots \\ X_n \end{pmatrix} \tag{7.126}$$

Now let us define the matrix

$$A = \begin{pmatrix} a_{11} & \cdots & a_{1n} \\ \vdots & \ddots & \vdots \\ a_{n1} & \cdots & a_{nn} \end{pmatrix}, \qquad a_{ij} = a_{ji} \tag{7.127}$$

Then A is a symmetric matrix. The quadratic form $Q(X)$ is a homogeneous second-degree expression in n variables and is defined by

$$Q(X) = \mathbf{X}^T A \mathbf{X} \tag{7.128}$$

where $\mathbf{X}^T$ is the transpose of **X**. If a quadratic form with real coefficients has the property that it is equal or greater than zero for all real values of its variables it is said to be positive. A positive form which is zero only for the values $X_1 = X_2 = \ldots = X_n = 0$ is said to be positive definite.

7.5.3 Theorem

A necessary and sufficient condition that the real quadratic form $\mathbf{X}^T A \mathbf{X}$ be positive definite is that every principal minor of A be positive.

7.6 Problems

7.1: Show that (7.51) and (7.56) are identical by performing the required matrix multiplications.

7.2: Show in detail the steps between (7.98) and (7.102).

7.3: Prove that the flux parts in the phenomenological equations (7.88) which involve the Casimir coefficients run reversibly. Hint: Determine $(TQ_s)_S$ and show that the sum of the Casimir flux parts does not include any dissipation expressions.

7.4: Ignoring superposition effects but including the kinetic energy of diffusion, the diffusion flux for the liquid water can be written as

$$\mathbf{J}^2 = \rho^2 \mathbf{v}_{2,d} = -L^{22}\left(\nabla(\mu_2 - \mu_0)_T + \frac{d\mathbf{v}_{2,d}}{dt}\right)$$

We may interpret this equation as the equation of relative motion of a suspended water droplet of mass ϵ_2

$$\underbrace{\epsilon_2 \frac{d\mathbf{v}_{2,d}}{dt}}_{\text{inertial force}} + \underbrace{\frac{\rho^2 \epsilon_2}{L^{22}}\mathbf{v}_{2,d}}_{\text{frictional force}} = \underbrace{-\epsilon_2 \nabla(\mu_2 - \mu_1)_T}_{\text{external force}}$$

Reduce the equation to a one-dimensional equation of motion in the z-coordinate system and determine the relative droplet velocity $w_{2,d}$ as a function of time. The initial condition is $w_{2,d}(t = 0) = 0$.
You may make the following simplifying assumptions:

(i) $w_{2,d}$ depends on t only and not on z,
(ii) ignore ordinary diffusion in comparison to baro-diffusion,
(iii) $\mathbf{v}_2 < \mathbf{v}_0$,
(iv) the hydrostatic equation $\partial p/\partial z = -g\rho$ applies,
(v) $\rho^2/L^{22} = $ constant, $\epsilon_2 = $ constant.

7.5: Use (7.95) to investigate the vertical temperature structure in a mixture of moist air. The specific humidity is $q = m^1 = 1 - m^0$. From the kinetic gas theory we find approximately

$$\frac{L^{00}}{T} = K_1 T^{1/2}, \qquad L^{00}\left(\frac{\partial \mu_1}{\partial m^1}\right)_{p,T}\frac{1}{1 - m^1} = K_2 T^{3/2}$$

K_1 and K_2 are constants.

Find the vertical temperature distribution subject to the following conditions:

(i) horizontal homogeneity,
(ii) vanishing heat flux,
(iii) the hydrostatic equation $\partial p / \partial z = -g\rho$ applies,
(iv) $T_0 = T(z = 0)$–boundary temperature,
(v) q decreases linearly with height: $q = q_0 - Az$, $q_0 = q(z = 0)$, $A = $ constant.
(vi)

$$\left(\frac{\partial \mu_1}{\partial m^1} \right)_{p,T} \frac{1}{1 - m^1} \approx \frac{R_0 T}{m^1}$$

where R_0 is the gas constant of dry air.

7.6: In a mixture of moist air the diffusion flux of water vapor is given by (7.96). The specific humidity of the air is $q = m^1 = 1 - m^0$. Find the height distribution of q by assuming horizontal homogeneity of all state variables. The following conditions are assumed to apply:

(i) the atmosphere is isothermal,
(ii) the hydrostatic equation $\partial p / \partial z = -g\rho$ applies,
(iii) the water vapor flux vanishes,
(iv) the boundary value is $q(z = 0) = q_0$. You may use the following approximations:

$$\rho = \frac{p}{R_0 T}, \qquad \left(\frac{\partial \mu_1}{\partial m^1} \right)_{p,T} \frac{1}{1 - m^1} \approx \frac{R_0 T}{m^1},$$

where R_0 is the gas constant of dry air.

7.7: Consider the vertical diffusion of water vapor in air in the interval $(0, L)$ assuming the conditions

$$\nabla T = 0, \quad \nabla p = 0, \quad \rho = \text{constant}, \quad \frac{dm^1}{dt} \approx \frac{\partial m^1}{\partial t}, \quad D^{11} = \text{constant} > 0$$

Find $m^1(z, t)$ by assuming a trial solution of the form

$$m^1 = C \exp(\alpha z + \beta t)$$

where C, α, β are constants. To determine the free parameters α and β use the following conditions

(i) $m^1(z, t = 0) = f(z)$ where $f(z)$ is some function of height,
(ii) $m^1(z = 0, t) = 0$,
(iii) $m^1(L, t) = 0$,
(iv) $m^1(z, t \to \infty) = 0$.

8

State functions of ideal gases

8.1 Basic relationships

For practically all purposes, in the absence of clouds and aerosol particles, the atmospheric system may be treated as a mixture of ideal gases. This mixture consists of dry air and water vapor. The dry air is composed of nitrogen N_2, oxygen O_2 and a large number of trace gases. While the water vapor is highly variable in space and time, the composition of the dry air is constant at least in the lower 25 km of the atmosphere over the entire globe and over extended periods of time. The observed CO_2 increase has little influence on the conventional thermodynamics of the air which will be considered in the following sections. For this reason it is almost a practical necessity to treat the dry air mixture as a single gas so that the moist air is composed of only two ideal gases.

Every ideal gas in the pure phase (unmixed) satisfies the *ideal gas law*

$$\boxed{v_i = \mathring{v}_i = \frac{R_i T}{p}} \qquad (8.1)$$

where the subscript $_i$ refers to the individual gas i. R_i is the individual gas constant, p and T are the pressure and the temperature of the gas. Actually, to the symbol v_i the superscript $^\circ$ should be attached ($\mathring{v}_i$) to indicate the pure phase. At first glance this notation may appear too pedantic, but later this pedantry is needed when we deal with mixtures of ideal gases. Otherwise we will be hopelessly lost in our thermodynamic manipulations.

The *individual gas constant* can be easily calculated from the *universal gas constant R^** with the help of *Avogadro's law*. This law states that all ideal gases at arbitrary but fixed (p, T)-values occupy the same volume which is known as the *molar volume $\mathring{v}$*. Let N^i stand for the number of moles contained in mass M^i of molecular weight $\check{m}_i$. In the pure phase $\check{v}_i$ is defined by

$$\check{v}_i = \frac{V}{N^i} = \frac{V\check{m}_i}{M^i} = v_i\check{m}_i = \frac{R_i T}{p}\check{m}_i \tag{8.2}$$

The product $\check{m}_i R_i$ is independent of the gas type and is the well-known *universal gas constant* R^*

$$\boxed{R^* = \check{m}_i R_i = 8.31432 \quad \mathrm{J\,mole^{-1}\,K^{-1}}} \tag{8.3}$$

8.2 The ideal gas law for dry air

As previously stated, we will treat the dry air as a mixture of ideal gases. We will label each gas by the symbol i so that the concentration of the i-th component is given by

$$\overset{\circ}{m}{}^{i} = \frac{\overset{\circ}{M}{}^{i}}{M^0} \quad \text{with} \quad M^0 = \sum_{i=4}^{n+4} \overset{\circ}{M}{}^{i} \tag{8.4}$$

where M^0 is the total mass of the dry air and the superscript $^\circ$ of the term $\overset{\circ}{M}{}^{i}$ refers to the pure (unmixed) phase. The summation must be started with $i = 4$ since the superscripts 1, 2 and 3 refer to water vapor, liquid water and ice.

Applying the general relationship $\psi = \psi_n m^n$ to the dry air, we obtain for the *specific volume of dry air* (subscript $_0$)

$$v_0 = \sum_{i=4}^{n+4} \overset{\circ}{m}{}^{i} \check{v}_i = \sum_{i=4}^{n+4} \overset{\circ}{m}{}^{i} \frac{R_i T}{p^0} \tag{8.5a}$$

We are tempted to write p_0 instead of p^0 but we must refrain from doing this since p_0 is reserved as a reference pressure usually taken as $p_0 = 1000\,\mathrm{hPa}$. If the dry air were the only gas present we could have omitted super- and subscripts altogether. Equation (8.5a) motivates us to introduce the *gas constant for dry air* R_0,

$$\boxed{p^0 v_0 = T \sum_{i=4}^{n+4} \overset{\circ}{m}{}^{i} R_i = R_0 T \quad \text{with} \quad R_0 = \sum_{i=4}^{n+4} \overset{\circ}{m}{}^{i} R_i} \tag{8.5b}$$

Since both $\overset{\circ}{m}{}^{i}$ and R_i are constants, R_0 is a constant also. Application of (8.3) shows that we can also define the gas constant for dry air by

$$R_0 = \frac{R^*}{\check{m}_0} \tag{8.6}$$

Table 8.1. *Data specifying the main components of dry air*

Gas	Molecular weight (g mole^{-1})	Mass fraction	Gas constant (J kg^{-1} K^{-1})	Molar fraction
N_2	28.013	0.7553	296.80	0.7809
O_2	31.999	0.2314	259.83	0.2095
Ar	39.948	0.0128	208.13	0.0093
CO_2	44.010	0.0005	188.92	0.0003

assuming that the mean molecular weight $\breve{m}_0$ of the dry air is known. Eliminating R_0 in (8.5b) and (8.6) and using (8.3) gives

$$\frac{1}{\breve{m}_0} = \sum_{i=4}^{n+4} \frac{\overset{\circ}{m}^i}{\breve{m}_i} \tag{8.7}$$

The mass concentrations of the individual component gases as well as the molecular weights are known quantities. Therefore, with the help of Table 8.1 we can easily obtain the mean molecular weight of the dry air as $\breve{m}_0 = 28.9644$ g mole^{-1}. Again using Table 8.1, from (8.5b) we find that $R_0 = 287.05$ J kg^{-1} K^{-1} which we also could have obtained directly from (8.6).

We conclude this section by stating some useful general relations which then will be applied to the dry air. If N^i stands for the number of moles contained in M^i, we have

$$N^i = \frac{M^i}{\breve{m}_i} \quad \text{with} \quad \sum_i M^i = M \tag{8.8a}$$

where the sum is taken over all partial masses. The *molar fraction* n^i is defined as

$$n^i = \frac{N^i}{N} \quad \text{with} \quad \sum_i N^i = N, \qquad \sum_i n^i = 1 \tag{8.8b}$$

where N is the number of moles contained in the total mass M. From (8.8a) follows

$$n^i = \frac{M^i}{N\breve{m}_i} \tag{8.8c}$$

so that

$$\frac{M}{N} = \breve{m} = \sum_i n^i \breve{m}_i \tag{8.8d}$$

Application of these formulas to the dry air of mass M^0 gives

$$\overset{\circ}{N}{}^i = \frac{\overset{\circ}{M}{}^i}{\breve{m}_i} \quad \text{with} \quad \sum_{i=4}^{n+4} \overset{\circ}{M}{}^i = M^0$$

$$\overset{\circ}{n}{}^i = \frac{\overset{\circ}{N}{}^i}{N^0} \quad \text{with} \quad \sum_{i=4}^{n+4} \overset{\circ}{N}{}^i = N^0 \tag{8.9}$$

$$\breve{m}_0 = \sum_{i=4}^{n+4} \overset{\circ}{n}{}^i \breve{m}_i$$

We should note that the gas constant for the dry air R_0 is obtained by weighting the individual gas constants R_i with the concentration $\overset{\circ}{m}{}^i$ while the mean molecular weight $\breve{m}_0$ of the dry air is found by weighting the individual molecular weights $\breve{m}_i$ with the molar fractions $\overset{\circ}{n}{}^i$. Dividing $\overset{\circ}{N}{}^i$ in (8.9) by M^0 results in

$$\frac{\overset{\circ}{N}{}^i}{M^0} = \frac{\overset{\circ}{m}{}^i}{\breve{m}_i} = \overset{\circ}{n}{}^i \frac{N^0}{M^0} = \frac{\overset{\circ}{n}{}^i}{\breve{m}_0} \tag{8.10}$$

Since

$$\frac{R_i}{R_0} = \frac{\breve{m}_0}{\breve{m}_i} \tag{8.11}$$

we obtain the useful relationship

$$\overset{\circ}{m}{}^i R_i = \overset{\circ}{n}{}^i R_0 \tag{8.12}$$

8.3 Dalton's law

Consider a mixture of ideal gases. Each individual gas will contribute to the total pressure of the mixture. The *partial pressure* p^i of the i-th gas is defined as the pressure the gas would have if it existed alone at the given temperature T occupying the volume of the mixture. Analogously, v^i is the *partial specific volume* the gas would have if it existed alone at the given temperature and at the pressure of the mixture. Dalton's law for the mixture of ideal gases may be expressed by

$$\boxed{p = \sum_i p^i, \qquad v = \sum_i v^i} \tag{8.13}$$

We now apply (8.13) to the mixture of the dry air with total pressure $p = p^0$ and total specific volume v_0. Reference to equation (8.5a) gives

$$\text{(a)} \quad p^0 = \sum_{i=4}^{n+4} \overset{\circ}{m}{}^i R_i \frac{T}{v_0} \quad \text{with} \quad p^i = \overset{\circ}{m}{}^i R_i \frac{T}{v_0}$$

$$\tag{8.14}$$

$$\text{(b)} \quad v_0 = \sum_{i=4}^{n+4} \overset{\circ}{m}{}^i R_i \frac{T}{p^0} \quad \text{with} \quad v^i = \overset{\circ}{m}{}^i R_i \frac{T}{p^0}$$

Table 8.2. *Molar fraction of various trace gases*

Gas	Molar fraction
Ne	1.8×10^{-5}
He	5.2×10^{-6}
CH_4	1.5×10^{-6}
CO	1.0×10^{-7}
O_3	$\leq 10^{-5}$, variable

Comparison of (8.14b) with (8.5a) shows a very useful relation between the partial volume and the pure phase specific volume as expressed by

$$v^i = \overset{\circ}{m}{}^i \overset{\circ}{v}_i \tag{8.15}$$

Since $\overset{\circ}{m}{}^i R_i = \overset{\circ}{n}{}^i R_0$, see (8.12), we obtain with the help of (8.5b)

$$p^i = \overset{\circ}{n}{}^i p^0, \qquad v^i = \overset{\circ}{n}{}^i v_0 \tag{8.16}$$

Utilizing (8.11) and (8.5b) the molar fraction $\overset{\circ}{n}{}^i = v^i/v_0$, which is also called the *volume fraction*, can also be expressed as

$$\overset{\circ}{n}{}^i = \overset{\circ}{m}{}^i \frac{\check{m}_0}{\check{m}_i} \tag{8.17}$$

Since the mass concentrations and the molecular weights are known, the final column in Table 8.1 can be easily calculated. The molar fractions of various atmospheric trace gases are listed in Table 8.2. It is seen that these gases have such a low concentration that they can be neglected in our thermodynamic considerations. However, in connection with air chemistry, these small concentrations might be significant.

8.4 Potentials of the ideal gas in the pure phase

The derivation of the state functions in the pure phase requires only the knowledge of the ideal gas law and of the specific heats. The pure phase is completely characterized by two of the state variables v, p, T. In mixtures the concentrations must also be known to specify the system. We will now derive various expressions for the potentials of the individual ideal gases. These expressions will then be generalized to deal with moist air. Finally, in a later chapter, the potentials for cloudy air will also be obtained to describe the thermodynamic processes in clouds.

The starting point of the analysis is *Gibbs' fundamental equation* for the pure phase ($dm^k = 0$). Replacing p/T from the ideal gas law, we obtain

$$d\mathring{s}_k = \frac{d\mathring{e}_k}{T} + d(R_k \ln \mathring{v}_k) \tag{8.18}$$

8.4.1 Internal energy

In Section 4.5 we have shown that the internal energy of an ideal gas is a function of temperature only so that

$$d\mathring{e}_k = \left(\frac{\partial \mathring{e}_k}{\partial T}\right)_{\mathring{v}_k} dT = c_{v,k}\, dT \tag{8.19a}$$

Integration of this expression gives

$$\mathring{e}_k(T) = \mathring{e}_k(T_0) + \int_{T_0}^{T} c_{v,k}(T')\, dT' \tag{8.19b}$$

If $c_{v,k}$ is treated as a constant, the integration can be carried out immediately. In general, we may write

$$\delta\mathring{e}_k = c_{v,k}\delta T \tag{8.19c}$$

Any form of (8.19) is known as the *caloric equation*. We should remark that the internal energy cannot be completely determined because of the constant value $\mathring{e}_k(T_0)$, T_0 is an arbitrary reference temperature usually taken as the melting point $T_0 = 273.15\,\mathrm{K}$. Whenever a reference point for pressure is needed we select $p_0 = 1000\,\mathrm{hPa}$. In all prognostic equations only differential expressions of the potentials occur so that the unknown reference values do not appear.

Reference to (3.55b) and (3.60) shows that for an ideal gas with $\mathring{e}_k = \mathring{e}_k(T)$ the *specific heat at constant* pressure is given by

$$c_{p,k} = c_{v,k} + p\left(\frac{\partial \mathring{v}_k}{\partial T}\right)_p \tag{8.20}$$

yielding the well-known relation

$$\boxed{c_{p,k}(T) - c_{v,k}(T) = R_k} \tag{8.21}$$

which differs from gas to gas.

8.4.2 Enthalpy

Substituting the ideal gas law into the definition of the enthalpy, we obtain

$$\mathring{h}_k(T) = \mathring{e}_k(T) + p\mathring{v}_k = \mathring{e}_k(T) + R_k T \tag{8.22}$$

showing that the enthalpy of an ideal gas is a function of temperature only. Introducing the definition for $c_{p,k}$ according to (3.43a) we find

$$d\mathring{h}_k = \left(\frac{\partial \mathring{h}_k}{\partial T}\right)_p dT = c_{p,k}\, dT \tag{8.23a}$$

Integration gives

$$\mathring{h}_k(T) = \mathring{h}_k(T_0) + \int_{T_0}^{T} c_{p,k}(T')\, dT' \tag{8.23b}$$

The general form is

$$\delta \mathring{h}_k = c_{p,k} \delta T \tag{8.23c}$$

8.4.3 Entropy

Repeating Gibbs' fundamental equation (8.18) we may write

$$d\mathring{s}_k = \frac{d\mathring{h}_k}{T} - \frac{\mathring{v}_k}{T} dp = \frac{d\mathring{e}_k}{T} + \frac{p}{T} d\mathring{v}_k \tag{8.24}$$

With the definitions $d\mathring{e}_k = c_{v,k} dT$, $d\mathring{h}_k = c_{p,k} dT$ and the ideal gas law we have

$$d\mathring{s}_k = c_{p,k}\frac{dT}{T} - \frac{R_k}{p} dp = c_{v,k}\frac{dT}{T} + \frac{R_k}{\mathring{v}_k} d\mathring{v}_k \tag{8.25}$$

Integration of these expressions gives

$$\mathring{s}_k(p, T) = \mathring{s}_k(p_0, T_0) + \int_{T_0}^{T} c_{p,k}(T')\, d\ln T' - R_k \ln \frac{p}{p_0} \tag{8.26a}$$

$$\mathring{s}_k(v_k, T) = \mathring{s}_k(v_k(0), T_0) + \int_{T_0}^{T} c_{v,k}(T')\, d\ln T' + R_k \ln \frac{\mathring{v}_k}{v_k(0)} \tag{8.26b}$$

with $v_k(0) = \mathring{v}_k(p_0, T_0)$ so that $\mathring{s}_k(p_0, T_0) = \mathring{s}_k(v_k(0), T_0)$. As in the case of the other potentials, the integration can be carried out analytically if $c_{p,k}$ and $c_{v,k}$ are treated as constants. The general form is given by

$$\delta \mathring{s}_k = c_{p,k} \delta \ln T - R_k \delta \ln p = c_{v,k} \delta \ln T + R_k \delta \ln \mathring{v}_k \tag{8.26c}$$

In order to have a concise and simple notation, we will separate one part of the equation which depends on temperature only from the remaining equation. Later, all partial differentiations with temperature held constant will be particularly simple. Therefore, we write in place of (8.26)

(a) $\overset{\circ}{s}_k(p, T) = \overset{+}{s}_k(T) - R_k \ln p$ with

$$\overset{+}{s}_k(T) = \overset{\circ}{s}_k(p_0, T_0) + \int_{T_0}^{T} c_{p,k}(T') \, d\ln T' + R_k \ln p_0$$

(b) $\overset{\circ}{s}_k(v_k, T) = \overset{+}{s}_k(T) + R_k \ln \overset{\circ}{v}_k$ with   (8.27)

$$\overset{+}{s}_k(T) = \overset{\circ}{s}_k(v_k(0), T_0) + \int_{T_0}^{T} c_{v,k}(T') \, d\ln T' - R_k \ln \overset{\circ}{v}_k(0)$$

(c) $\overset{+}{s}_k(T) - \overset{+}{s}_k(T) = R_k \ln(R_k T)$ since $p_0 \overset{\circ}{v}_k(0) = R_k T_0$

At this point it may be worthwhile to point out the relationship between entropy and the *potential temperature* θ which is often used in meteorological analysis[1]

$$\boxed{\theta = T\left(\frac{p_0}{p}\right)^{R_0/c_{p,0}}, \qquad p_0 = 1000 \text{ hPa}} \qquad (8.28)$$

In variational form we may write

$$\delta \ln \theta = \delta \ln T - \frac{R_0}{c_{p,0}} \delta \ln p \qquad (8.29)$$

Comparison with (8.26c) shows that

$$\delta \overset{\circ}{s}_0 = c_{p,0} \delta \ln \theta \qquad (8.30)$$

The quantities R_0, $c_{p,0}$ and p refer to dry air. Later (8.28) will be derived as part of a more complex system.

8.4.4 Chemical potential

The chemical potential is defined as

$$\overset{\circ}{\mu}_k = \overset{\circ}{h}_k - T\overset{\circ}{s}_k \qquad (8.31)$$

Again we will separate a part which depends on temperature only. According to (8.23) the enthalpy depends only on temperature, the specific entropy (8.27a), however, depends on both temperature and pressure. Substituting these two equations into (8.31) gives

$$\overset{\circ}{\mu}_k(p, T) = \overset{\circ}{h}_k(T) - T(\overset{+}{s}_k(T) - R_k \ln p) = \overset{+}{\mu}_k(T) + R_k T \ln p$$

$$\text{with} \quad \overset{+}{\mu}_k(T) = \overset{\circ}{h}_k(T) - T\overset{+}{s}_k(T) \qquad (8.32)$$

---

[1] A detailed derivation of the equation for the potential temperature will be given in Section 12.1.

where $\overset{+}{\mu}_k$ has precisely the form of the Gibbs potential. Inspection of (8.32) shows that it was not possible, in contrast to the entropy, to separate the temperature dependency completely.

8.4.5 Free energy

The free energy is defined by

$$\overset{\circ}{f}_k = \overset{\circ}{e}_k - T\overset{\circ}{s}_k = \overset{\circ}{\mu}_k - \overset{\circ}{v}_k p = \overset{\circ}{\mu}_k - R_k T \tag{8.33}$$

where use has been made of the ideal gas law. All we need to do is to replace the chemical potential from (8.32) to obtain

$$\overset{\circ}{f}_k(p, T) = \overset{+}{\mu}_k(T) + R_k T \ln p - R_k T = \overset{+}{f}_k(T) + R_k T \ln p$$
$$\text{with} \quad \overset{+}{f}_k(T) = \overset{+}{\mu}_k(T) - R_k T \tag{8.34a}$$

where $\overset{+}{f}_k(T)$ has precisely the form of the free energy. Again, it was not possible to separate the temperature dependency completely since the remaining part of the equation depends on temperature also.

We will now proceed to obtain an expression for $\overset{\circ}{f}_k$ as function of the temperature and the specific volume instead of temperature and pressure. In this case we make use of (8.27b) and find

$$\overset{\circ}{f}_k(v_k, T) = \overset{\circ}{e}_k(T) - T(\overset{+}{s}_k(T) + R_k \ln \overset{\circ}{v}_k) = \overset{++}{f}_k(T) - R_k T \ln \overset{\circ}{v}_k$$
$$\text{with} \quad \overset{++}{f}_k(T) = \overset{\circ}{e}_k(T) - T\overset{+}{s}_k(T) \tag{8.34b}$$

Again it is not possible to separate the temperature dependency completely.

8.5 Potentials of moist air

8.5.1 Derivation of the ideal gas law for moist air

A very helpful application of the free energy potential is the derivation of the gas law for moist air, which we consider as an ideal mixture of dry air and water vapor. As a useful exercise we will re-derive the gas law for the dry air but this time using the free energy concept. The extension to moist air is then a fairly simple matter. In a later chapter we will then generalize the atmospheric system to include clouds.

8.5.1.1 The ideal gas law for dry air

We consider the dry air as a mixture of n ideal gases. The *ideal mixture* is defined by the following property:

if F_i represents the free energy of the component i of the mixture, then the sum of the individual components

$$F = \sum_i F_i \tag{8.35}$$

is equal to the free energy of the mixture. The F_i are so formed as if the i-th component by itself fills the volume of the mixture at temperature T. According to (8.34b) the quantity F_i is given by

$$F_i = f_i \overset{\circ}{M}{}^i = \overset{\circ}{M}{}^i \left(\overset{+}{f}_i(T) - R_i T \ln \frac{V}{\overset{\circ}{M}{}^i} \right) \tag{8.36}$$

With $V/\overset{\circ}{M}{}^i = v/\overset{\circ}{m}{}^i$ we find for the dry air mixture

$$\overset{\circ}{f}_0(v, T) = \sum_{i=4}^{n+4} \overset{\circ}{m}{}^i \left[\overset{+}{f}_i(T) + R_i T (\ln \overset{\circ}{m}{}^i - \ln v) \right] \tag{8.37}$$

The ideal gas law can be found with the help of the equation (see Table 6.1)

$$p = -\left(\frac{\partial \overset{\circ}{f}_0}{\partial v} \right)_{T, \overset{\circ}{m}{}^i} = \sum_{i=4}^{n+4} \overset{\circ}{m}{}^i R_i \frac{T}{v} \tag{8.38}$$

or

$$pv = R_0 T \quad \text{with} \quad R_0 = \sum_{i=4}^{n+4} \overset{\circ}{m}{}^i R_i \tag{8.39}$$

It will be a useful exercise to give the formal justification of why it is possible to treat the dry air mixture as a pure phase gas. To show this, we rewrite (8.37) with the help of (8.15) and (8.16). Combining the logarithms leads to the fraction

$$\frac{v}{\overset{\circ}{m}{}^i} = \frac{v_0}{\overset{\circ}{m}{}^i} = v_0 \frac{\overset{\circ}{v}_i}{v^i} = \frac{\overset{\circ}{v}_i}{\overset{\circ}{n}{}^i} \tag{8.40}$$

so that we obtain

$$\overset{\circ}{f}_0(v, T) = \sum_{i=4}^{n+4} \overset{\circ}{m}{}^i (\overset{+}{f}_i(T) - R_i T \ln \overset{\circ}{v}_i + R_i T \ln \overset{\circ}{n}{}^i) \tag{8.41}$$

Comparison with (8.34b) shows that the pure phase of the i-th component of the mixture is given by

$$\overset{\circ}{f}_i = \overset{+}{f}_i(T) - R_i T \ln \overset{\circ}{v}_i \tag{8.42}$$

so that

$$\overset{\circ}{f}_0(v, T) = \sum_{i=4}^{n+4} \overset{\circ}{m}{}^i \overset{\circ}{f}_i \tag{8.43}$$

with

$$f_i = \overset{\circ}{f}_i + R_i T \ln \overset{\circ}{n}{}^i \tag{8.44}$$

Thus, the free energy of the i-th component differs from the pure phase by $R_i T \ln \overset{\circ}{n}{}^i$ which is known as the *mixing term*. Introducing into (8.37) the gas constant for dry air according to (8.5b), we may also write for the free energy of the dry air the expression

$$\overset{\circ}{f}_0(v, T) = \overset{+}{f}_0(T) - R_0 T \ln v_0 \quad \text{with}$$

$$\overset{+}{f}_0(T) = \sum_{i=4}^{n+4} \overset{\circ}{m}{}^i \left[\overset{+}{f}_i(T) + R_i T \ln \overset{\circ}{m}{}^i \right] \tag{8.45}$$

By comparing this expression with (8.34b) it is seen that the equation for the free energy of the dry air mixture has the same form as the corresponding equation for the free energy of the pure phase. This is the formal justification of treating the mixture of the dry air as if it were a pure phase gas.

8.5.1.2 The ideal gas law for moist air

The free energy of the moist air F_m can be found from a simple extension of (8.36) by including the contribution of water vapor F_1. Recognizing that now the volume refers to moist air, we may write

$$F_m = \sum_{i=4}^{n+4} F_i + F_1 = \sum_{i=4}^{n+4} M^i \left(\overset{+}{f}_i(T) - R_i T \ln \frac{V_m}{\overset{\circ}{M}{}^i} \right) + M^1 \left(\overset{+}{f}_1(T) - R_1(T) \ln \frac{V_m}{M^1} \right) \tag{8.46}$$

where V_m and M^1 represent the volume of moist air and the mass of water vapor. The quantity $\overset{+}{f}_1(T)$ can be found from (8.34b) since that formula refers to any pure phase gas. Using the definitions as stated next,

$$\sum_{i=4}^{n+4} \overset{\circ}{M}{}^i = M^0, \qquad M^0 + M^1 = M_m, \qquad \frac{M^0}{M_m} = \overset{0}{m}_m, \qquad \frac{M^1}{M_m} = m^1_m,$$

$$\frac{\overset{\circ}{M}{}^i}{M^0} = \overset{\circ}{m}{}^i, \qquad \frac{\overset{\circ}{M}{}^i}{M_m} = \overset{\circ}{m}{}^i m^0_m, \qquad f_m = \frac{F_m}{M_m}$$

$$V_m = v_m M_m = v M_m, \qquad \frac{V_m}{\overset{\circ}{M}{}^i} = \frac{v M_m}{\overset{\circ}{M}{}^i} = \frac{v}{\overset{\circ}{m}{}^i m^0_m} \tag{8.47}$$

equation (8.46) can be immediately rewritten in the useful form

$$f_m(v, T) = \sum_{i=4}^{n+4} \overset{\circ}{m}{}^i m^0_m \left[\overset{+}{f}_i(T) - R_i T \ln \frac{v}{\overset{\circ}{m}{}^i m^0_m} \right] + m^1_m \left[\overset{+}{f}_1(T) - R_1 T \ln \frac{v}{m^1_m} \right] \tag{8.48}$$

In order to obtain the ideal gas law for moist air, we proceed as in the case of dry air and obtain from (8.48)

$$p = -\left(\frac{\partial f}{\partial v}\right)_{T, m_m^k, \mathring{m}^i;} = \left(m_m^0 \sum_{i=4}^{n+4} \mathring{m}^i R_i + m_m^1 R_1\right)\frac{T}{v} \qquad (8.49)$$

Using (8.5b),

$$pv = \left(m_m^0 R_0 + m_m^1 R_1\right)T = R_m T \quad \text{with} \quad R_m = m_m^0 R_0 + m_m^1 R_1 \qquad (8.50)$$

we have found the gas law for moist air where R_m is the corresponding gas constant.

In reality, the quantity R_m is not a constant at all since m_m^0, m_m^1 change as the atmospheric water vapor concentration changes. It is somewhat awkward to use R_m. This motivates us to put the product $R_m T$ in a practical form. Since

$$R_1 = \frac{\breve{m}_0}{\breve{m}_1} R_0 = 1.60789 R_0 \approx 1.61 R_0 \qquad (8.51)$$

we may write

$$R_m T = R_0(1 + 0.61q)\,T \qquad (8.52)$$

where we have replaced m_m^1 by the more common symbol q denoting the *specific humidity*. The fictitious temperature

$$\boxed{T_v = (1 + 0.61q)\,T} \qquad (8.53a)$$

is called the *virtual temperature* of the moist air. The quantity

$$\Delta T_v = T_v - T = 0.61q\,T \qquad (8.53b)$$

is known as the *virtual temperature correction*, which is some measure of humidity. At the end of this chapter we will present some additional humidity measures. With (8.53a) the gas law for moist air assumes the common form

$$\boxed{pv = R_0 T_v} \qquad (8.54)$$

where p is the total pressure made up by the partial pressures of the dry air and of the water vapor. Obviously, the specific volume refers to moist air. A comparison of the ideal gas laws for dry and moist air permits a physical description of the virtual temperature. This, by definition, is the temperature of dry air having the same pressure and specific volume as the moist air as can be recognized from $T_v = pvT/p^0 v_0$.

8.5.1.3 Partial pressures of moist air

Denoting the partial pressure (specific volume) of the dry air and the water vapor by p^0 and p^1 (v^0 and v^1), we can write from (8.50)

$$p = p^0 + p^1 = \left(m_m^0 R_0 + m_m^1 R_1\right)\frac{T}{v} = R_m\frac{T}{v}$$
$$v = v^0 + v^1 = \left(m_m^0 R_0 + m_m^1 R_1\right)\frac{T}{p} = R_m\frac{T}{p} \tag{8.55}$$

and identify

$$p^k = m_m^k R_k \frac{T}{v}, \qquad v^k = m_m^k R_k \frac{T}{p}, \qquad k = 0, 1 \tag{8.56}$$

Since $T/v = p/R_m$ and $T/p = v/R_m$ we may also write these expressions in a form analogous to (8.16),

$$p^k = \frac{m_m^k R_k}{m_m^0 R_0 + m_m^1 R_1}p, \qquad v^k = \frac{m_m^k R_k}{m_m^0 R_0 + m_m^1 R_1}v, \qquad k = 0, 1 \tag{8.57}$$

Substituting $R_k = R^*/\breve{m}_k$ into (8.57) gives immediately

$$p^k = \frac{m_m^k/\breve{m}_k}{m_m^0/\breve{m}_0 + m_m^1/\breve{m}_1}p, \qquad v^k = \frac{m_m^k/\breve{m}_k}{m_m^0/\breve{m}_0 + m_m^1/\breve{m}_1}v, \qquad k = 0, 1 \tag{8.58}$$

Using the relations (8.8a) and (8.8b) for number of moles and molar fraction, we obtain

$$N^k = \frac{M^k}{\breve{m}_k}, \qquad n^k = \frac{N^k}{N_m}, \qquad \frac{m_m^k}{\breve{m}_k} = \frac{n^k N_m}{M_m}, \qquad k = 0, 1 \tag{8.59}$$

and then from (8.58)

$$p^k = n^k p, \qquad v^k = n^k v, \qquad k = 0, 1 \tag{8.60}$$

Finally, we get from (8.56) the important conversion relation with $v = v_m$

$$\frac{v_m}{m_m^k} = \frac{R_k T}{p^k} = \frac{R_k T}{n^k p} = \frac{\breve{v}_k}{n^k}, \qquad k = 0, 1 \tag{8.61}$$

This conversion relation permits us to rewrite the free energy equation (8.48) for the moist air in a form permitting the deduction of the remaining potentials.

8.5.2 Free energy of moist air

Substituting (8.61) into the free energy expression for the moist air (8.48) gives

$$
\begin{aligned}
f_m = m_m^0 &\left[\sum_{i=4}^{n+4} \overset{+}{m}^i (\overset{+}{f}_i(T) + R_i T \ln \overset{+}{m}^i) - R_0 T \ln \overset{\circ}{v}_0 + R_0 T \ln n^0 \right] \\
&+ m_m^1 \left[\overset{+}{f}_1(T) - R_1 T \ln \overset{\circ}{v}_1 + R_1 T \ln n^1 \right] \\
= f_0 m_m^0 &+ f_1 m_m^1
\end{aligned}
\tag{8.62}
$$

where use has been made of (8.5b). Comparison with (8.34b) and (8.37) (where now the specific volume of dry air must be written as $\overset{\circ}{v}_0$) shows that

$$
f_k = \overset{\circ}{f}_k + R_k T \ln n^k, \qquad k = 0, 1
\tag{8.63}
$$

Thus, it is seen that the free energy of component k differs from the pure phase by the mixing term $R_k T \ln n^k$. Furthermore, from the definition of the pure phase (8.34b) and from (8.61) follows

$$
\overset{\circ}{f}_k = \overset{+}{f}_k(T) - R_k T \ln \overset{\circ}{v}_k = \overset{+}{f}_k(T) - R_k T \ln \frac{v_m}{m_m^k} - R_k T \ln n^k, \qquad k = 0, 1
\tag{8.64}
$$

Therefore, with (8.63) we can write

$$
\boxed{f_k = \overset{+}{f}_k(T) - R_k T \ln \frac{v_m}{m_m^k}, \qquad k = 0, 1}
\tag{8.65}
$$

With the expressions we have just derived we are now in the position to obtain the remaining potentials. By now the reader, hopefully, is convinced that the somewhat complex notation we have used is absolutely necessary to prevent us from making mistakes.

8.5.3 Entropy of moist air

According to Table 6.1 we can write for the entropy of moist air

$$
s = s_m = - \left(\frac{\partial f_m}{\partial T} \right)_{v, m_m^k} = - \left(\frac{\partial f_0}{\partial T} \right)_v m_m^0 - \left(\frac{\partial f_1}{\partial T} \right)_v m_m^1
\tag{8.66}
$$

From (8.63) we get directly the required partial derivatives

$$
\left(\frac{\partial f_k}{\partial T} \right)_v = \left(\frac{\partial \overset{\circ}{f}_k}{\partial T} \right)_{\overset{\circ}{v}_k} + R_k \ln n^k, \qquad k = 0, 1
\tag{8.67}
$$

Substituting this expression into the previous equation, we obtain for the entropy of the moist air

$$s_m = -\left[\left(\frac{\partial \mathring{f}_0}{\partial T}\right)_{\mathring{v}_0} + R_0 \ln n^0\right] m_m^0 - \left[\left(\frac{\partial \mathring{f}_1}{\partial T}\right)_{\mathring{v}_1} + R_1 \ln n^1\right] m_m^1 \qquad (8.68)$$

In this latter equation we now have partial derivatives ($k = 0, 1$) involving the specific free energy of the pure phase. We will now eliminate these in terms of expressions already available. With the help of (8.61) we find from (8.65)

$$\left(\frac{\partial f_k}{\partial T}\right)_v = \frac{d\overset{+}{f}_k(T)}{dT} - R_k \ln \frac{\mathring{v}_k}{n^k}, \qquad k = 0, 1 \qquad (8.69)$$

Combining (8.69) with (8.67) gives for the partial derivative of the pure phase

$$\left(\frac{\partial \mathring{f}_k}{\partial T}\right)_{\mathring{v}_k} = -\mathring{s}_k = \frac{d\overset{+}{f}_k(T)}{dT} - R_k \ln \mathring{v}_k, \qquad k = 0, 1 \qquad (8.70)$$

The function $\overset{+}{f}_k(T)$ has been defined in connection with (8.34b). We are now ready to state the final form of the entropy equation for the moist air from (8.68).

$$s_m = s_0 m_m^0 + s_1 m_m^1 = (\mathring{s}_0 - R_0 \ln n^0) m_m^0 + (\mathring{s}_1 - R_1 \ln n^1) m_m^1 \qquad (8.71)$$

Therefore,

$$\boxed{s_k = \mathring{s}_k - R_k \ln n^k \quad \text{with} \quad \mathring{s}_k = \overset{+}{s}_k(T) - R_k \ln p, \qquad k = 0, 1} \qquad (8.72)$$

where $\mathring{s}_k$ was taken from (8.27a). As in the case of the free energy the potential s_k differs from the pure phase $\mathring{s}_k$ by a mixing term.

8.5.4 Chemical potential of moist air

This was previously defined by

$$\mu = \mu_0 m_m^0 + \mu_1 m_m^1 = f + pv = f_0 m_m^0 + f_1 m_m^1 + p\left(v_0 m_m^0 + v_1 m_m^1\right) \qquad (8.73)$$

where we have expanded f and v. The question now arises quite naturally if the partial specific volume is equal to the pure phase or if it possesses a mixing term. Reference to (8.50) and (8.1) shows that

$$v = v_0 m_m^0 + v_1 m_m^1 = \frac{R_0 T}{p} m_m^0 + \frac{R_1 T}{p} m_m^1 = \mathring{v}_0 m_m^0 + \mathring{v}_1 m_m^1 \qquad (8.74)$$

so that $v_k = \mathring{v}_k$, i.e. the specific partial volume is equal to the pure phase. Substituting now the definition of the f_k from (8.63) into (8.73) we find

$$\mu = \mu_0 m_m^0 + \mu_1 m_m^1 = (\mathring{f}_0 + R_0 T \ln n^0 + p\mathring{v}_0) m_m^0 + (\mathring{f}_1 + R_1 T \ln n^1 + p\mathring{v}_1) m_m^1 \tag{8.75}$$

Therefore, we may identify

$$\mathring{\mu}_k = \mathring{f}_k + p\mathring{v}_k = \overset{+}{f}_k(T) + R_k T \ln p + p\mathring{v}_k = \overset{+}{\mu}_k(T) + R_k T \ln p, \qquad k = 0, 1 \tag{8.76}$$

where we have used (8.34a). Finally, we find for the chemical potential the expression

$$\boxed{\mu_k = \overset{+}{\mu}_k(T) + R_k T (\ln n^k + \ln p), \qquad k = 0, 1} \tag{8.77}$$

The chemical potential differs from the pure phase by the mixing term $R_k T \ln n^k$. Equation (8.77) was previously used with $k = 1$ to deal with the concept of baro-diffusion.

8.5.5 Internal energy of moist air

The internal energy can be defined in terms of the potentials of the free energy and the entropy. For these potentials expressions are available which can be directly substituted into

$$e = e_0 m_m^0 + e_1 m_m^1 = f + Ts = f_0 m_m^0 + f_1 m_m^1 + T \left(s_0 m_m^0 + s_1 m_m^1 \right) \tag{8.78}$$

Substituting from (8.63) and (8.72) into (8.78) we find

$$\begin{aligned} e = \mathring{e}_0 m_m^0 + \mathring{e}_1 m_m^1 &= (\mathring{f}_0 + R_0 T \ln n^0 + T\mathring{s}_0 - R_0 T \ln n^0) m_m^0 \\ &+ (\mathring{f}_1 + R_1 T \ln n^1 + T\mathring{s}_1 - R_1 T \ln n^1) m_m^1 \end{aligned} \tag{8.79}$$

As can be directly seen from this expression, the mixing terms of the free energy and the entropy cancels out. The expressions in parentheses have precisely the form (8.78) of the internal energy. Therefore, we may write

$$\boxed{e_k(T) = \mathring{e}_k(T) = \mathring{f}_k + T\mathring{s}_k, \qquad k = 0, 1} \tag{8.80}$$

Hence, the internal energy potential is equal to the potential of the pure phase. From (8.19b), treating the specific heat $c_{v,k}$ as a constant as is permissible for practically all meteorological applications, we find

$$\mathring{e}_k(T) = \mathring{e}_k(T_0) + c_{v,k}(T - T_0), \qquad k = 0, 1 \tag{8.81}$$

8.5.6 Enthalpy of moist air

The enthalpy can be defined in terms of the specific Gibbs function and the entropy by

$$h = h_0 m_m^0 + h_1 m_m^1 = \mu + Ts = \mu_0 m_m^0 + \mu_1 m_m^1 + T \left(s_0 m_m^0 + s_1 m_m^1 \right) \quad (8.82)$$

From (8.72) and (8.77) we recognize that the mixing terms of the chemical potential and of the specific partial entropy s_k are given by $R_k T \ln n^k$ and $-R_k \ln n^k$. Thus, the effects of the mixing terms cancel precisely since in (8.82) s_k is multiplied by the temperature T. Therefore, the enthalpy potential is equal to the potential of the pure phase. From (8.23) we find

$$\boxed{h_k = \overset{\circ}{h}_k \quad \text{with} \quad \overset{\circ}{h}_k = \overset{\circ}{h}_k(T) = \overset{\circ}{h}_k(T_0) + c_{p,k}(T - T_0), \qquad k = 0, 1} \quad (8.83)$$

if the specific heat $c_{p,k}$ is held constant.

8.6 Humidity measures

We conclude this chapter by stating some humidity measures for the moist air which do not involve the saturation or the chemical equilibrium. The remaining humidity measures such as the relative humidity and the wet bulb temperature will be treated when the proper relationships have been established.

8.6.1 Specific humidity

$$\boxed{m_m^1 = q = \frac{M^1}{M^0 + M^1} = \frac{M^1}{M_m}} \quad (8.84)$$

8.6.2 Mixing ratio

$$\boxed{r = \frac{M^1}{M^0}} \quad (8.85)$$

8.6.3 Conversion between q and r

$$q = \frac{r}{1 + r} \quad \text{or} \quad r = \frac{q}{1 - q} \quad (8.86)$$

In the atmosphere r and q vary between 0 and 0.05 so that $r \approx q$. It is customary in the meteorological practice to replace r or q by 1000 times their regular values.

For example, if $r = 0.01$, we then have $r = 10$ g of water vapor per kg of dry air. However, in all equations involving r or q the original definitions must be used.

8.6.4 Molar fraction

The molar fraction, as previously defined, can be expressed in terms of r or q as shown next.

$$n^1 = \frac{N^1}{N_m} = \frac{N^1}{N^0 + N^1} = \frac{M^1/\breve{m}_1}{M^0/\breve{m}_0 + M^1/\breve{m}_1} = \frac{M^1}{M^0 R_0/R_1 + M^1} = \frac{r}{R_0/R_1 + r}$$
(8.87)

Since r, q are much smaller than 1, we may also write

$$\boxed{n^1 \approx \frac{R_1}{R_0} r \approx \frac{R_1}{R_0} q}$$
(8.88)

8.6.5 Virtual temperature correction

The virtual temperature correction, as defined by (8.53b), can be expressed in terms of the molar fraction and is given by

$$\boxed{\Delta T_v = 0.61 q\, T \approx 0.61 \frac{R_0}{R_1} n^1 T}$$
(8.89)

8.6.6 Absolute humidity

Let ρ^1 denote the water vapor density in a mixture of dry air and water vapor. With M^1 the mass of the water vapor in the volume V of the mixture, the absolute humidity is defined by

$$\boxed{\rho^1 = \frac{M^1}{V}}$$
(8.90)

Usually this quantity is given in kg of water vapor per m^3.

8.7 Problems

8.1: Calculate the enthalpy change of an ideal gas undergoing a reversible adiabatic expansion so that $V_f = 5/3 V_i$ where the subscripts i and f denote the initial and final state. Assume that the heat capacities are constants. Let $C_p = 20.93$ J mole^{-1} K^{-1} and $T_i = 250$ K.
Hint: First show that $T_f = T_i(V_i/V_f)^{k-1}$ where $k = C_p/C_v$.

8.2: The specific Gibbs function (specific free enthalpy) of a substance is given by

$$\mu(T_0, p_0) = \mu(T_0, p_0) - (T - T_0)s(T_0, p_0) + c_p(T - T_0) - c_pT \ln\left(\frac{T}{T_0}\right)$$

$$+ RT \ln\left(\frac{p}{p_0}\right) + RTA_1(p - p_0) + \frac{RT}{2}A_2\left(p^2 - p_0^2\right)$$

The individual gas constant R, the specific heat c_p, and the quantities A_1, A_2 are constants. For the value at the reference point the relation $\mu(T_0, p_0) = h(T_0, p_0) - T_0s(T_0, p_0)$ is valid.

Find the specific entropy, the specific volume, and the specific enthalpy as functions of T and p.

8.3: The specific free energy of a substance is given by the expression

$$f(T, v) = f(T_0, v_0) - (T - T_0)s(T_0, v_0) + \int_{T_0}^{T} \beta(T')dT' - T\int_{T_0}^{T}\frac{\beta(T')}{T'}dT'$$

$$+ RT[Ei(a/(RTv)) - Ei(a/(RT_0v_0))] + R(T - T_0)\exp(a/(RT_0v_0))$$

$R, a, T_0, v_0, f(T_0, v_0)$ are constants. The function $\beta(T)$ is not specified. $Ei(x)$ is the first exponential integral with $d\,Ei(x)/dx = \exp(x)/x$. For the reference point we set $f(T_0, v_0) = e(T_0, v_0) - T_0s(T_0, v_0)$.

Find the presssure, the specific entropy, the specific internal energy, the specific enthalpy, the free enthalpy, and the specific heat c_v as a function of (T, v).

8.4: Consider a number of inert gases separated from one another by suitable partitions. All gases are at the same temperature and pressure. Box k contains gas A_k and N^k moles.

(a) Calculate the entropy of the system before the partitions are removed.
(b) Calculate the entropy of the system after the partitions are removed.
(c) Calculate the entropy change due to the diffusion of any number of inert gases.
(d) Calculate the diffusion of one mole of A_1 and one mole of A_2.

8.5: The free energy of a gas in natural coordinates is given by

$$f(T, v) = f(0) - (T - T_0)(s(0) - c_v) - Tc_v \ln\left(\frac{T}{T_0}\right)$$

$$- RT \ln\left(\frac{v - \alpha}{v_0 - \alpha}\right) - \beta\left(\frac{1}{v} - \frac{1}{v_0}\right)$$

with $f(0) = e(0) - T_0s(0)$. The quantities $T_0, v_0, e(0), s(0)$ are reference values while c_v, R, α, β are constants.

Find the specific entropy, the specific internal energy, the specific heat at constant pressure, and the pressure as functions of T and v.

9

State functions of the condensed pure phase

The purpose of this chapter is to derive expressions for the specific partial quantities $\psi_k = v_k, s_k, e_k, h_k, f_k, \mu_k, \; k = 2, 3$ for the condensed phases of water vapor. These quantities are needed to describe the thermodynamics of cloud air. In our treatment we assume that the liquid water and ice occur unmixed and neither liquid water nor ice contain foreign materials. Due to this assumption we cannot describe the formation of a water droplet. Such a droplet forms when water vapor condenses on a suitable aerosol particle so that the resulting droplet cannot be viewed as a pure substance. With this in mind, we may consider the ψ_k as the pure phase, i.e. $\psi_k = \overset{\circ}{\psi}_k(p, T)$. For simplicity we leave out the superscript $^\circ$ denoting the pure phase and also drop the suffix $_k$ since confusion is unlikely. In the final section we will add the suffix $_k$ for completeness and accuracy.

9.1 The material coefficients

First of all, we need to define the coefficients of the *isothermal compressibility* (κ) and the *adiabatic compressibility* (κ_s). They are defined by

$$\kappa = -\frac{1}{v}\left(\frac{\partial v}{\partial p}\right)_T \geq 0 \qquad (9.1)$$

$$\kappa_s = -\frac{1}{v}\left(\frac{\partial v}{\partial p}\right)_s \geq 0 \qquad (9.2)$$

Similarly, we define the *isochoric pressure coefficient* β and the *isobaric expansion coefficient* v^* by

$$\beta = \frac{1}{p}\left(\frac{\partial p}{\partial T}\right)_v \qquad (9.3)$$

$$v^* = \frac{1}{v}\left(\frac{\partial v}{\partial T}\right)_p \qquad (9.4)$$

137

The quantities κ, β and v^* are not independent. We recognize this by writing the state equation according to the Appendix of Chapter 6 in the form

$$f(v, p, T) = 0 \qquad (9.5)$$

Expansion of (9.5) gives

$$df = \left(\frac{\partial f}{\partial p}\right)_{v,T} dp + \left(\frac{\partial f}{\partial v}\right)_{p,T} dv + \left(\frac{\partial f}{\partial T}\right)_{v,p} dT = 0 \qquad (9.6)$$

This expression is also valid if we choose any of the independent variables T, v and p as constant. Thus, we may write down the expressions

$$dT = 0: \qquad \left(\frac{\partial v}{\partial p}\right)_{T} = -\left(\frac{\partial f}{\partial p}\right)_{v,T} \left(\frac{\partial f}{\partial v}\right)_{p,T}^{-1}$$

$$dv = 0: \qquad \left(\frac{\partial p}{\partial T}\right)_{v} = -\left(\frac{\partial f}{\partial T}\right)_{v,p} \left(\frac{\partial f}{\partial p}\right)_{v,T}^{-1} \qquad (9.7)$$

$$dp = 0: \qquad \left(\frac{\partial v}{\partial T}\right)_{p} = -\left(\frac{\partial f}{\partial T}\right)_{v,p} \left(\frac{\partial f}{\partial v}\right)_{p,T}^{-1}$$

Furthermore, we easily recognize that

$$\left(\frac{\partial v}{\partial T}\right)_{p} \left(\frac{\partial T}{\partial p}\right)_{v} \left(\frac{\partial p}{\partial v}\right)_{T} = -1 \qquad (9.8)$$

Substituting (9.7) into (9.1), (9.3) and (9.4) we find the required relation

$$p\beta\kappa = v^* \qquad (9.9)$$

showing that β, κ, v^* are not independent so that any one of these can be expressed by the two remaining coefficients. Relation (9.8) is quite general and was derived solely from the equation of state (9.5).

For the ideal gas in the pure phase all material coefficients can be easily determined from the ideal gas law. For the calculation of κ_s it is also necessary to apply Gibbs' fundamental equation. The result is

$$\kappa = \frac{1}{p}, \qquad \kappa_s = -\frac{c_v}{c_p p}, \qquad \beta = \frac{1}{T}, \qquad v^* = \frac{1}{T} \qquad (9.10)$$

We will now return to the condensed pure phase and briefly discuss the definition of the *incompressibility*. If $\kappa = 0$ then $(\delta v)_T = 0$, $-$ pressure incompressibility.

If $v^* = 0$ then $(\delta v)_p = 0$. Therefore, when both v^* and κ are zero then $\delta v = 0$ and $v = $ constant or the material is incompressible.

In case of mixed systems the situation is more complicated since the equation of state assumes the form $v = v(p, T, m^j)$ where the m^j refer to the substances of impurity. A salt solution of varying concentrations is incompressible for all practical purposes, however, it is not homogeneous. Now thermodynamic relations must also include concentration gradients, thus increasing the complexity of the thermodynamic system. In our discussion we will exclude these complexities. Furthermore, to simplify the notation, in this section we will omit the subscript $k = 2, 3$.

9.2 The thermal equation of state

For the range of pressures and temperatures of the terrestrial atmosphere we may safely assume that v^* and κ are constants. To obtain the desired equation of state we think of equation (9.5) as being solved in the form $v = v(p, T)$. Expanding this expression gives

$$dv = \left(\frac{\partial v}{\partial T}\right)_p dT + \left(\frac{\partial v}{\partial p}\right)_T dp = v(v^* dT - \kappa dp) \qquad (9.11)$$

The solution of this differential equation can be written down at once,

$$v = v(0)\exp\left[v^*(T - T_0) - \kappa(p - p_0)\right] \quad \text{with} \quad v(0) = v(p_0, T_0) \qquad (9.12)$$

Separating the part depending on temperature only, we may write

$$\overset{+}{v}(T) = v(0)\exp\left[v^*(T - T_0)\right] \qquad (9.13)$$

Since the argument of the exponential function $\exp[-\kappa(p - p_0)]$ is a very small number, we may discontinue its expansion after the linear term. Thus we obtain the approximate expression

$$\boxed{v = \overset{+}{v}(T)[1 - \kappa(p - p_0)]} \qquad (9.14)$$

This expression will be needed soon to find simplified equations for the potentials of the condensed pure phase.

9.3 Potentials of the condensed pure phase

In order to calculate the potentials we need some information, in addition to the thermal equation of state, about the dependency of the specific heat on temperature

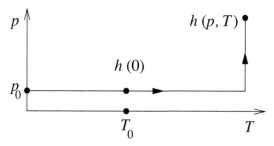

Fig. 9.1 Special path of integration.

and pressure. For the heat capacity c_p we may state the identity

$$c_p(p, T) = c_p(p_0, T) + \int_{p_0}^{p} \left(\frac{\partial c_p}{\partial p'} \right)_T dp' \tag{9.15}$$

We assume that $c_p(p_0, T)$ is a known function. We will now proceed to formulate the various potentials and solve the volume integral.

9.3.1 Enthalpy

For convenience we repeat some definitions from (3.43) involving the enthalpy. In the pure phase ($m^2 = 1$ or $m^3 = 1$) we may write

$$\left(\frac{\partial h}{\partial T} \right)_p = c_p(p, T), \qquad \left(\frac{\partial h}{\partial p} \right)_T = \gamma + v \tag{9.16}$$

According to (6.58) the quantity γ is defined by

$$\gamma = -T \left(\frac{\partial v}{\partial T} \right)_p \tag{9.17}$$

so that with (9.4) we find

$$\gamma = -v v^* T \tag{9.18}$$

Expanding $h(p, T)$, using the above identity, gives

$$dh = \left(\frac{\partial h}{\partial T} \right)_p dT + \left(\frac{\partial h}{\partial p} \right)_T dp = c_p(p, T)dT + v(1 - v^* T)dp \tag{9.19}$$

Since the integration between two points involving the variable of state h is independent of the path, we may use the special path shown in Figure 9.1 so that

$$\boxed{h(p, T) = h(0) + \int_{p_0,T_0}^{p_0,T} c_p dT' + (1 - v^* T) \int_{p_0,T}^{p,T} v dp'} \tag{9.20}$$

As previously stated we assume that $c_p(p_0, T)$ is known.

9.3.2 Entropy

We repeat from (6.55) the following two identities

$$\left(\frac{\partial s}{\partial T}\right)_p = \frac{c_p}{T}, \qquad \left(\frac{\partial s}{\partial p}\right)_T = \frac{\gamma}{T} = -vv^* \tag{9.21}$$

Expanding $s(p, T)$ we obtain

$$ds = \left(\frac{\partial s}{\partial T}\right)_p dT + \left(\frac{\partial s}{\partial p}\right)_T dp = \frac{c_p}{T}dT - vv^* dp \tag{9.22}$$

Using the integration path of Figure 9.1 we easily find

$$s(p, T) = s(0) + \int_{p_0,T_0}^{p_0,T} c_p d\ln T' - v^* \int_{p_0,T}^{p,T} v dp' \tag{9.23}$$

9.3.3 Chemical potential

All that needs to be done is to substitute (9.20) and (9.23) into the definition $\mu = h - Ts$. This gives

$$\mu(p, T) = h(0) - Ts(0) + \int_{p_0,T_0}^{p_0,T} c_p dT' - T\int_{p_0,T_0}^{p_0,T} c_p d\ln T' + \int_{p_0,T}^{p,T} v dp' \tag{9.24}$$

To have a simplified notation we introduce the abbreviations

$$\mu(0) = h(0) - T_0 s(0)$$

$$\overset{+}{\mu}(T) = \mu(0) - (T - T_0)s(0) + \int_{p_0,T_0}^{p_0,T} c_p dT' - T\int_{p_0,T_0}^{p_0,T} c_p d\ln T' \tag{9.25a}$$

and find

$$\mu(p, T) = \overset{+}{\mu}(T) + \int_{p_0,T}^{p,T} v dp' \tag{9.25b}$$

9.3.4 Internal energy and free energy

All that needs to be done to obtain the proper expressions is to substitute (9.20) and (9.25b) into the definitions for e and f

$$e(p, T) = h(p, T) - pv(p, T)$$
$$f(p, T) = \mu(p, T) - pv(p, T) \tag{9.26}$$

If desired, the resulting expressions for e and f can be easily written down. We will omit this exercise.

In order to handle the various potentials we must evaluate the volume integral. The evaluation is rather simple since we treat the compressibility κ as a constant. Using the approximation (9.14) we find with excellent approximation

$$\int_{p_0,T}^{p,T} v \, dp' = v(0) \exp\left[v^*(T - T_0)\right](p - p_0)\left[1 - \frac{\kappa}{2}(p - p_0)\right] \qquad (9.27)$$

In the integration we could have used (9.12) instead of (9.14) and would have obtained a more exact expression but the advantage would be very slim. With the availability of (9.27) the potentials are properly specified.

9.4 The simplified state functions

In atmospheric systems the condensed phases of water vapor may be treated as incompressible. Therefore, with more than sufficient accuracy, we may write

$$v^* = 0, \qquad \kappa = 0 \qquad (9.28)$$

Since we are dealing with incompressible materials, there is no need to make a distinction between specific heat at constant volume or pressure. Therefore, we drop the suffix p and simply write c_k instead of $c_{p,k}$. The c_k may be temperature dependent but we will treat them as constants. With this assumption we obtain without difficulty the following expression

$$\int_{p_0,T}^{p,T} v \, dp' = v(0)(p - p_0) \qquad (9.29)$$

and, therefore,

$$\mu(p, T) = \overset{+}{\mu}(T) + v(0)(p - p_0) \qquad (9.30)$$

In the previous sections we have dropped the suffix k for simplicity. At this point we will add this suffix for completeness and accuracy. Summarizing, we obtain the following expressions for the condensed phases of liquid water ($k = 2$) and ice ($k = 3$).

$$v_k^* = 0, \qquad \kappa_k = 0, \qquad k = 2, 3 \qquad (9.31)$$

$$v_k = v_k(0) = \text{constant}_k, \qquad k = 2, 3 \qquad (9.32)$$

$$s_k(T) = s_k(0) + c_k \ln \frac{T}{T_0}, \qquad k = 2, 3 \qquad (9.33)$$

$$h_k(p, T) = h_k(0) + c_k(T - T_0) + v_k(p - p_0), \qquad k = 2, 3 \qquad (9.34)$$

$$e_k(T) = h_k - pv_k = e_k(0) + c_k(T - T_0), \qquad k = 2, 3 \qquad (9.35)$$

$$\mu_k(p, T) = \overset{+}{\mu}_k(T) + v_k(p - p_0) \quad \text{with}$$

$$\overset{+}{\mu}_k(T) = \mu_k(0) - (T - T_0)s_k(0) + c_k(T - T_0) - Tc_k \ln \frac{T}{T_0}, \qquad k = 2, 3 \qquad (9.36)$$

$$f_k(T) = \overset{+}{\mu}_k(T) - p_0 v_k, \qquad k = 2, 3 \qquad (9.37)$$

In Chapters 8 and 9 we have assembled most of the information required to formulate the thermodynamics of cloudy air.

9.5 Problems

9.1: Verify equation (9.10).

9.2: Verify the following expressions:

$$\gamma = -Tvv^*, \qquad \delta = Tp\beta, \qquad c_p - c_v = \frac{Tv(v^*)^2}{k}$$

$$\left(\frac{\partial c_p}{\partial p}\right)_T = -Tv\left[\left(\frac{\partial v^*}{\partial T}\right)_p + (v^*)^2\right], \qquad \left(\frac{\partial c_v}{\partial v}\right)_T = Tp\left[\left(\frac{\partial \beta}{\partial T}\right)_v + \beta^2\right]$$

10

State functions for cloud air

In the previous two chapters we have presented the state functions for moist air and for the condensed phase. We are now ready to present the state functions for clouds consisting of dry air, water vapor, liquid water and ice particles. The general formation law is repeated from (1.16),

$$\psi = \sum_{k=0}^{3} \psi_k m^k = \psi_n m^n, \qquad \sum_{k=0}^{3} m^k = 1 \tag{10.1}$$

where $\psi = v, s, e, h, f, \mu$. The concentrations m^k add up to 1 so that we may eliminate any one of these, say m^i, and obtain

$$\psi = \psi_i + \sum_{k=0}^{3} (\psi_k - \psi_i) m^k \tag{10.2}$$

Since we are dealing with cloud air it will be of advantage to introduce the total water concentration by means of

$$m^{H_2O} = m^1 + m^2 + m^3 \tag{10.3}$$

Substituting this expression into (10.2) gives

$$\psi = \psi_0 m^0 + \psi_i m^{H_2O} + \sum_{k=1}^{3} (\psi_k - \psi_i) m^k, \quad i = 0, \ldots, 3 \tag{10.4}$$

Often it is permissible to simplify the atmospheric system by neglecting v_2, v_3, in comparison to v_0, v_1, which differ by several orders of magnitude. Therefore, we often set approximately $v_2 = v_3 = 0$ but we retain the liquid water and ice concentrations ($m^2 \neq 0, m^3 \neq 0$). Let the superscript ^{wi} stand for 'water and ice', i.e. the condensed phase. To obtain a compact notation we define the following quantities

$$M = M^m + M^{wi}, \qquad M^m = M^0 + M^1, \qquad M^{wi} = M^2 + M^3$$

$$m^m = m^0 + m^1, \qquad m^{wi} = m^2 + m^3, \qquad v^m = \frac{V^m}{M} = v_0 m^0 + v_1 m^1$$

$$v^{wi} = \frac{V^{wi}}{M} = v_2 m^2 + v_3 m^3, \qquad v_m = \frac{V^m}{M^m} = \frac{v^m}{m^m}, \qquad v_{wi} = \frac{V^{wi}}{M^{wi}} = \frac{v^{wi}}{m^{wi}}$$

$$(10.5)$$

These definitions are possible since cloud water and ice occur unmixed. Whenever necessary, we will show that the approximation $v_2 = v_3 = 0$ has been used by writing the corresponding expressions with the symbol $\approx$ instead of $=$. One way to consistently introduce this approximation is to simplify a potential such as (9.36)

$$\mu_k = \overset{+}{\mu}_k(T) + (p - p_0)v_k \implies \mu_k \approx \overset{+}{\mu}_k(T), \qquad k = 2, 3 \qquad (10.6)$$

10.1 Specific volume and potentials

10.1.1 Specific volume

According to (8.1), (8.50), (10.1), and (10.5) we obtain for the specific volume

$$v = \sum_{k=0}^{3} v_k m^k = (R_0 m^0 + R_1 m^1)\frac{T}{p} + v^{wi}$$

$$= R_m(m^0 + m^1)\frac{T}{p} + v^{wi} = R_m m^m \frac{T}{p} + v^{wi} = v^m + v^{wi}$$

$$(10.7)$$

yielding

$$p = \frac{R_m m^m T}{v - v^{wi}} = \frac{R_m m^m T}{v^m} = \frac{R_m T}{v_m} \qquad (10.8)$$

For many practical purposes it is permissible to ignore the liquid water and ice concentrations ($m^2, m^3 \ll 1$) so that $m^m \approx 1$. Therefore,

$$v \approx \frac{R_m T}{p} \approx v^m \approx v_m \qquad (10.9)$$

10.1.2 Internal energy

It is a simple matter to substitute the specific partial values e_k into the general formation law. With the help of (8.81) and (9.35) we can write

$$e = e_n m^n = e_n(0)m^n + c_{v,n}m^n(T - T_0) \quad \text{with} \quad e_k(0) = \overset{\circ}{e}_k(T_0) \qquad (10.10a)$$

where we have assumed that all specific heats are independent of temperature. Again using the general formation law, we find a compact form for the internal energy as given by

$$e = e(T_0) + c_v(T - T_0), \qquad e(T_0) = e_n(T_0)m^n$$
$$c_v = c_{v,n}m^n, \qquad c_{v,k} = c_k, \qquad k = 2, 3 \tag{10.10b}$$

We observe that $e(T, m^k)$ does not explicitly depend on p and v so that $\delta = p$ in atmospheric systems, see (3.55).

10.1.3 Enthalpy

Analogous to (10.10a) we write the general formation law for the enthalpy. Using equations (8.83) and (9.34) assuming that the specific heat is independent of temperature we find

$$h = h_n m^n = h_n(0)m^n + c_{p,n}m^n(T - T_0) + (v_2 m^2 + v_3 m^3)(p - p_0)$$
$$\text{with} \quad h_k(0) = \overset{o}{h}_k(p_0, T_0)$$
$$\tag{10.11a}$$

Applying the general formation law to c_p itself we finally obtain

$$h = h(0) + c_p(T - T_0) + (v_2 m^2 + v_3 m^3)(p - p_0)$$
$$h(0) = h_n(0)m^n, \qquad c_p = c_{p,n}m^n, \qquad c_{p,k} = c_k, \qquad k = 2, 3 \tag{10.11b}$$

Inspection of this equation shows that $h = h(p, T, m^k)$. Using (10.11b) and (3.43b) we find a useful expression for the specific heat of tension

$$\gamma = \left(\frac{\partial h}{\partial p}\right)_{T,m^k} - v = -(v - v^{wi}) = -v^m = -v_m m^m$$
$$\gamma = -R_m m^m \frac{T}{p} = -(R_0 m^0 + R_1 m^1)\frac{T}{p} \tag{10.12a}$$

Approximate but for many purposes adequate expressions for h and γ are obtained by ignoring the specific volumes of liquid water and ice. For the specific heat of tension we find

$$\gamma \approx -v \tag{10.12b}$$

and for the enthalpy

$$h \approx h(0) + c_p(T - T_0) \tag{10.13}$$

Before concluding this section, we will state still another but very useful expression for the enthalpy by applying (10.4) for $i = 1$ and (3.43c, d). The resulting expression for the enthalpy of cloud air is given by

$$h = h_0 m^0 + h_1 m^{H_2O} - l_{21} m^2 - l_{31} m^3$$
$$= h_0(0)m^0 + h_1(0)m^{H_2O} + (c_{p,0}m^0 + c_{p,1}m^{H_2O})(T - T_0) - l_{21}m^2 - l_{31}m^3$$

$$(10.14)$$

10.1.4 Entropy

In order to express the general formation law for the entropy we recall s_k from (8.72) and (8.27) for the moist air and from (9.33) for the condensed phase. Again assuming that the specific heat is temperature independent we find for the components of the moist air

$$s_k = \overset{+}{s}_k(T) - R_k \ln(n^k p) = \overset{\circ}{s}_k(0) + c_{p,k} \ln \frac{T}{T_0} - R_k \ln \frac{p^k}{p_0} \quad \text{with}$$

$$p^k = n^k p, \qquad \overset{+}{s}_k(T) = \overset{\circ}{s}_k(0) + c_{p,k} \ln \frac{T}{T_0} + R_k \ln p_0, \qquad k = 0, 1$$

$$(10.15)$$

The general formation law can then be written as

$$s = s(0) + \left(c_{p,0} \ln \frac{T}{T_0} - R_0 \ln \frac{p^0}{p_0} \right) m^0$$
$$+ \left(c_{p,1} \ln \frac{T}{T_0} - R_1 \ln \frac{p^1}{p_0} \right) m^1 + (c_2 m^2 + c_3 m^3) \ln \frac{T}{T_0}$$

$$(10.16)$$

with $s(0) = \overset{\circ}{s}_n(0)m^n, \qquad \overset{\circ}{s}_k(0) = \overset{\circ}{s}_k(p_0, T_0)$

Finally, using (10.4) for $i = 1$, we may write

$$s = s_0 m^0 + s_1 m^{H_2O} - \frac{l_{21} + a_{21}}{T} m^2 - \frac{l_{31} + a_{31}}{T} m^3 \qquad (10.17)$$

where use was made of the identities listed in (6.55). To obtain a really useful expression for the specific entropy s we substitute as the last step (10.15) into (10.17) and find

$$s = \overset{\circ}{s}_0(0)m^0 + \overset{\circ}{s}_1(0)m^{H_2O} + (c_{p,0}m^0 + c_{p,1}m^{H_2O}) \ln \frac{T}{T_0}$$
$$- R_0 m^0 \ln \frac{p^0}{p_0} - R_1 m^{H_2O} \ln \frac{p^1}{p_0} - \frac{l_{21} + a_{21}}{T} m^2 - \frac{l_{31} + a_{31}}{T} m^3$$

$$(10.18)$$

We would like to remark that we could have used $i = 2$ or $i = 3$ instead of $i = 1$ in (10.4) without changing the physical meaning. We then would have found equations analogous to (10.14) and (10.18).

10.1.5 Chemical potential

The general formation law according to (1.16) is given by

$$\mu = \mu_n m^n \tag{10.19}$$

The chemical potentials for the moist air components are found from (8.77)

$$\mu_k = \overset{+}{\mu}_k(T) + R_k T \ln p^k, \qquad k = 0, 1$$

$$\overset{+}{\mu}_k(T) = h_k(0) + c_{p,k}(T - T_0) - T\left(\overset{\circ}{s}_k(0) + c_{p,k} \ln \frac{T}{T_0} + R_k \ln p_0\right) \tag{10.20}$$

where use has been made of *Dalton's law* for partial pressures. For practical purposes a part $\overset{+}{\mu}_k(T)$ has been separated which depends on temperature only. This results in very compact expressions and makes partial differentiations very simple when T is to be held constant.

For the condensed phase the chemical potentials are given by (9.36),

$$\mu_k = h_k(0) - T s_k(0) + c_k(T - T_0) - T c_k \ln \frac{T}{T_0} + v_k(p - p_0) \tag{10.21}$$

$$\text{since} \quad \mu_k(0) = h_k(0) - T_0 s_k(0), \qquad k = 2, 3$$

Application of (10.19) finally gives

$$\mu = \left(h_0(0) - T\overset{\circ}{s}_0(0) + c_{p,0}(T - T_0) - T c_{p,0} \ln \frac{T}{T_0} + R_0 T \ln \frac{p^0}{p_0}\right) m^0$$

$$+ \left(h_1(0) - T\overset{\circ}{s}_1(0) + c_{p,1}(T - T_0) - T c_{p,1} \ln \frac{T}{T_0} + R_1 T \ln \frac{p^1}{p_0}\right) m^1$$

$$+ \left(h_2(0) - T\overset{\circ}{s}_2(0) + c_2(T - T_0) - T c_2 \ln \frac{T}{T_0} + v_2(p - p_0)\right) m^2$$

$$+ \left(h_3(0) - T\overset{\circ}{s}_3(0) + c_3(T - T_0) - T c_3 \ln \frac{T}{T_0} + v_3(p - p_0)\right) m^3 \tag{10.22}$$

The mathematical structure of this lengthy expression is very simple.

10.2 Phase changes

In this section we will consider coexisting phases of water in thermodynamic equilibrium. Often a distinction is made between *lower* and *higher phases*. The transformation from the lower to the higher phase requires the addition of heat. In this context the water vapor is in a higher phase than liquid water since the heat of vaporization must be added to transform the liquid water to water vapor. Similarly, liquid water is in a higher phase than ice.

We shall now examine the conditions of coexistence of the various phases. Let us consider an isolated system which is defined by the restriction that no exchange of mass and energy with the external surroundings takes place.

In Section 6.3.1 we found that the following conditions must exist:

$$dS = 0, \quad \text{restrictive conditions:} \quad dE = 0, \quad\quad dV = 0 \quad\quad (10.23)$$

We must add the condition that the total mass cannot change. Phase changes between the various M^k, $k = 1, 2, 3$ must be accounted for. With these conditions Gibbs' fundamental equation in extensive form

$$T dS = dE + p dV - \mu_n d M^n \quad\quad (10.24)$$

then reduces to

$$\text{(a)} \quad \mu_n d M^n = 0$$
$$\text{(b)} \quad \mu_n d m^n = 0 \quad\quad\quad (10.25)$$
$$\text{(c)} \quad \mu_n d N^n = 0$$

The equilibrium condition (10.25c) can be obtained from (10.25a) since the molecular weight is the same for all three phases.

We now consider the most simple case which is the coexistence between two phases only. One of these we take as the water vapor so that $dm^1 = -dm^k$, $k = 2$ or $k = 3$. From (10.25b) then follows

$$\mu_1(p, T) = \mu_k(p, T) \quad \text{or} \quad a_{k1}(p, T) = 0 \quad\quad (10.26)$$

The fact that μ_1 is a function only of (p, T) is a consequence of (1.24) since now $q = 1$. Equation (10.26) is precisely the condition of *chemical equilibrium* as introduced in Chapter 4 which indicates that there is a functional relation between the vapor pressure p and temperature T,

$$p = p(T) \quad\quad (10.27)$$

A plot of this relation gives the *saturation vapor pressure* curve as described by the Clausius–Clapeyron equation which will be derived soon.

10.3 Gibbs' phase rule

In the previous section we considered the equilibrium conditions of a heterogeneous system consisting of only one substance (water). We will now consider the equilibrium conditions of an isolated heterogeneous system consisting of more than one substance. Such a system is sometimes called a *multi-component system*. Chemical reactions of the various substances are not admitted.

As a matter of convention, we would like to point out that not only the physical states of a substance are called gas phase, liquid phase and solid phase, but one even speaks of different phases of a solid substance (sulphur, carbon) when differences occur in its internal structure.

Let us now consider a system consisting of r substances. The number of coexisting phases is v. The lowest phase is denoted by $'$, the next higher phase by $''$. The highest phase is v. The internal energy E, the entropy S and the volume V of the system are given by

$$E = E' + E'' + \cdots + E^v$$
$$S = S' + S'' + \cdots + S^v \qquad (10.28)$$
$$V = V' + V'' + \cdots + V^v$$

For the heterogeneous multi-component system the equilibrium condition is given (see Chapter 6) by

$$dS = dS' + dS'' + \cdots + dS^v = 0 \qquad (10.29a)$$

The restrictive conditions are

$$dV = dV' + dV'' + \cdots + dV^v = 0 \qquad (10.29b)$$

$$dE = dE' + dE'' + \cdots + dE^v = 0 \qquad (10.29c)$$

Since the mass of the isolated system remains unchanged, it will be necessary to include additional restrictions for the r substances $M_1, M_2, \ldots, M_r$ which may occur in any of the phases. We have written subscripts in order to avoid confusion with the masses of the water substance M^k, $k = 1, 2, 3$. These additional restrictive conditions are

$$dM_1' + \quad \cdots \quad + dM_1^v = 0$$
$$\vdots \qquad \vdots \qquad \vdots \qquad (10.30)$$
$$dM_r' + \quad \cdots \quad + dM_r^v = 0$$

Each phase by itself is a multi-component system so that Gibbs' fundamental equation must be valid for any one of the v phases. For the lowest phase we can write

$$dE' = T'dS' - p'dV' + \sum_{i=1}^{r} \mu_i' dM_i' \qquad (10.31)$$

Analogous equations can be written for the remaining phases. If these equations are added according to (10.29c), we obtain

$$T'dS' + T''dS'' + \cdots + T^v dS^v - p'dV' - p''dV'' - \cdots - p^v dV^v$$
$$+ \sum_{i=1}^{r} \left(\mu_i' dM_i' + \mu_i'' dM_i'' + \cdots + \mu_i^v dM_i^v \right) = 0 \qquad (10.32a)$$

In this equation we now eliminate dS', dV' and dM_i' with the help of (10.29a, b) and (10.30). The result is

$$(T'' - T')\,dS'' + \cdots + (T^\nu - T')\,dS^\nu - (p'' - p')\,dV'' - \cdots - (p^\nu - p')\,dV^\nu$$

$$+ \sum_{i=1}^{r} \left[\left(\mu_i'' - \mu_i' \right) dM_i'' + \cdots + \left(\mu_i^\nu - \mu_i' \right) dM_i^\nu \right] = 0$$

$$(10.32b)$$

We require that this equation be valid for arbitrary values of the independent differentials $d(S, V, M_i)$. This is possible only if the coefficients of the differentials vanish, i.e.

$$T' = T'' = \cdots = T^\nu = T$$
$$p' = p'' = \cdots = p^\nu = p \qquad\qquad (10.33)$$
$$\mu_i' = \mu_i'' = \cdots = \mu_i^\nu = \mu_i, \qquad i = 1, \ldots, r$$

The conclusion is that a heterogeneous multi-component system is in *thermodynamic equilibrium* if it is characterized by uniform temperature and pressure. Moreover, the chemical potentials of the various components must be the same for all phases. Written in implicit form, these quantities are related by equations

$$\psi_j(p, T, \mu_1, \mu_2, \ldots, \mu_r) = 0, \qquad j = 1, 2, \ldots, \nu \qquad (10.34)$$

There are $(r + 2)$ variables: $p, T, \mu_1, \ldots, \mu_r$. To prevent the equations in $(r + 2)$ unknowns to be overspecified, we must have

$$\nu \leq r + 2 \qquad\qquad (10.35a)$$

This is Gibbs' phase rule stating that in equilibrium the number of coexisting phases of a multi-component system can exceed the number of independent substances by no more than 2. The quantity

$$\boxed{f_d = r + 2 - \nu} \qquad\qquad (10.35b)$$

is known as the number of *degrees of freedom* of the system.

Two examples will clarify the situation. In the case of a one-component system with two phases we have $r = 1$ and $\nu = 2$, $f_d = 1$. The single degree of freedom describes the state of a system along a curve such as the saturation vapor pressure curve. In the case of a one-component system and three phases ($r = 1$, $\nu = 3$) we obtain $f_d = 0$. The system has no degree of freedom which is possible only at the *triple point*.

In the case of chemical reactions taking place which were excluded in the previous discussion, the number of degrees of freedom is reduced by the number R of independent reactions so that $f_d = r + 2 - \nu - R$.

10.4 Clausius–Clapeyron equations

Let us now consider in detail the stable equilibrium between two phases of water. In perfect equilibrium phase changes between the two phases cannot occur. This situation is characterized by vanishing affinities. First we consider the classical Clausius–Clapeyron equation which ignores the presence of dry air. As a next step we will take the pressure effect of the dry air into account but the advantage gained is very small.

10.4.1 Classical Clausius–Clapeyron equation

In this case the presence of dry air is ignored so that the molar fraction of water vapor $n^1 = 1$. Then water vapor, liquid water and ice are in the pure phase. This means that the chemical potentials depend only on temperature and pressure so that the Clausius–Clapeyron differential equations must be a relationship between pressure and temperature as already mentioned in the previous section.

The equilibrium condition for water vapor and liquid water or ice is given by

$$a_{k1}(p, T) = \mu_k - \mu_1 = 0, \qquad da_{k1}(p, T) = 0, \implies$$

$$\left(\frac{\partial a_{k1}}{\partial T}\right)_p dT + \left(\frac{\partial a_{k1}}{\partial p}\right)_T dp = 0, \qquad k = 2, 3 \tag{10.36}$$

Applying the Kelvin–De Donder relations (6.58b) to (10.36) we find

$$\frac{l_{k1}}{T} dT + (v_k - v_1) dp = 0 \quad \text{or}$$

$$\left(\frac{dp}{dT}\right)_{a_{k1}=0} = \frac{l_{k1}}{(v_1 - v_k)T}, \qquad k = 2, 3 \tag{10.37}$$

Using the identity $a_{32} = (\mu_3 - \mu_1) - (\mu_2 - \mu_1) = a_{31} - a_{21}$ and $l_{32} = l_{31} - l_{21}$ we obtain analogously to (10.37)

$$\left(\frac{dp}{dT}\right)_{a_{32}=0} = \frac{l_{32}}{(v_2 - v_3)T} \tag{10.38}$$

In order to solve the Clausius–Clapeyron equations (10.37) and (10.38), we must know the behavior of the latent heat as a function of the variables p and T. Recalling the relationships (3.43), (8.83) and (9.34), we can write

$$l_{k1} = h_1 - h_k, \qquad k = 2, 3 \quad \text{with}$$

$$h_1 = h_1(0) + c_{p,1}(T - T_0) \tag{10.39a}$$

$$h_k = h_k(0) + c_k(T - T_0) + v_k(p - p_0)$$

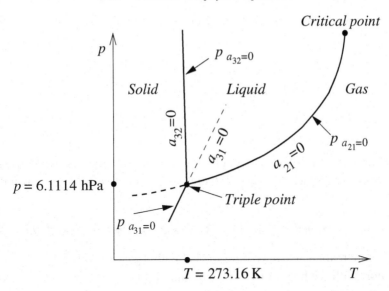

Fig. 10.1 Equilibrium curves as function of temperature for two phases of water.

and, therefore,

$$l_{k1} = l_{k1}(0) - (c_k - c_{p,1})(T - T_0) + v_k(p - p_0)$$

$$\text{with}\quad l_{k1}(0) = h_1(0) - h_k(0)$$

(10.39b)

To obtain an analytic solution to (10.37), with an acceptable approximation we set in (10.37) and (10.39b) $v_k = 0,\ k = 2, 3$. Thus we obtain

(a) $$\left(\frac{d\ln p}{dT}\right)_{a_{k1}=0} \approx \frac{l_{k1}}{R_1 T^2} \approx \frac{l_{k1}(0) - (c_k - c_{p,1})(T - T_0)}{R_1 T^2}$$

(b) $$\left(\frac{dp}{dT}\right)_{a_{32}=0} = \frac{l_{32}}{(v_2 - v_3)T} \approx \frac{l_{32}(0) - (c_3 - c_2)(T - T_0)}{(v_2 - v_3)T} \approx -10^5 \text{ hPa K}^{-1}$$

(10.40)

From the latter expression it is seen that in the (p, T)-plane the course of the equilibrium curve $p_{a_{32}=0}$ is almost parallel to the p-axis.

A schematic representation of the equilibrium curves is shown in Figure 10.1. The existence of the *triple point* was already demonstrated in connection with Gibbs' phase rule. The temperature and pressure of the triple point are given by $T = 273.16$ K and $p = 6.1114$ hPa. These values differ from those of the *ice point* with $T = 273.15$ K and $p = 1013.25$ hPa. Of no consequence to our work is the existence of the *critical point* ($p = 218\,000$ hPa, $T = 647$ K), where a distinction between the gas and the fluid is no longer possible, that is $v_1 = v_2$.

In Chapter 3 it was mentioned that the latent heat is not a variable of state. We will now give proof of this statement. Let P' and P'' refer to the lower and higher phase, respectively. The latent heat l for the transformation from P' to P'' is given by

$$l = \int_{P'}^{P''} đ(q + w) = \int_{P'}^{P''} (de + p\,dv) = \int_{P'}^{P''} dh = h'' - h'$$

$$= T \int_{P'}^{P''} ds = T(s'' - s')$$

(10.41)

since in equilibrium the pressure and the temperature are the same in both phases. $đ(q + w)$ is not an exact differential, therefore, l cannot be a variable of state.

10.4.2 Modified Clausius–Clapeyron equation for cloud air

In this more realistic case we have to deal with a gaseous mixture of dry air and water vapor. With the help of (8.77) we obtain

$$a_{k1}(p, T, n^1) = \mu_k(p, T) - \mu_1(p, T, n^1)$$
$$= \mu_k(p, T) - \mathring{\mu}_1(p, T) - R_1 T \ln n^1 = 0$$
$$\mu_1(p, T, n^1) = \mathring{\mu}_1(p, T) + R_1 T \ln n^1,$$
$$\mathring{\mu}_1(p, T) = \overset{+}{\mu}_1(T) + R_1 T \ln p, \quad k = 2, 3$$

(10.42)

Additional remarks on the a_{k1} will be given in the Appendix to this chapter. Instead of (10.37) we now get

$$\frac{l_{k1}}{T} dT + (v_k - v_1)\,dp - \frac{R_1 T}{n^1} dn^1 = 0 \quad \text{with}$$

$$v_1 = \frac{R_1 T}{p}, \qquad \frac{dp}{p} + \frac{dn^1}{n^1} = d \ln p^1, \qquad k = 2, 3$$

(10.43a)

Here p^1 is the partial pressure of water vapor. Equation (10.38), describing the equilibrium for water and ice ($a_{32} = 0$), remains unchanged. Setting $v_k = 0$ for $k = 2, 3$ as before we obtain

$$\frac{d \ln p^1}{dT} \approx \frac{l_{k1}}{R_1 T^2}, \qquad k = 2, 3$$

(10.43b)

Comparison of (10.43b) with (10.40a) shows that in the latter equation the pressure must be interpreted as the partial water vapor pressure p^1. As a matter of convention, from now on we denote the *saturation vapor pressure* by p^{k1}, i.e. $p^1(a_{k1} = 0) = p^{k1}$, $k = 2, 3$ and similarly $p(a_{32} = 0) = p^{32}$.

Finally, it should be observed that at the *boiling temperature* T_b of a fluid, the saturation vapor pressure is equal to the exterior pressure acting on the fluid. For water we have $T_b = 373.15$ K at $p = 1013.25$ hPa. If the exterior pressure is decreased, the boiling temperature is decreased also. For example, if the pressure is reduced $p = 10$ hPa, then the boiling temperature decreases to $T_b = 280$ K. If the pressure is increased to 10^5 hPa then the boiling temperature of water increases to 573 K.

10.5 Saturation vapor pressure curve, equilibrium curve for melting

Instead of directly integrating the Clausius–Clapeyron differential equation to obtain p^{kl} we will proceed in a different way leading, of course, to the same result. Setting the affinity equal to zero, we obtain from (10.42) with (9.36) and (8.77)

$$a_{k1} = \mu_k - \mu_1 = \overset{+}{\mu}_k(T) - \overset{+}{\mu}_1(T) + \left(\frac{p^{kl}}{n^1} - p_0\right)v_k - R_1 T \ln p^{kl} = 0, \quad k = 2, 3$$

$$(10.44)$$

In the absence of equilibrium, i.e. $a_{k1} \neq 0$, p^{kl} must be replaced by the partial pressure p^1. In order to find a suitable form for p^{kl} we write

$$\overset{+}{\mu}_k(T) - \overset{+}{\mu}_1(T) = R_1 T \ln \epsilon_{k1}, \quad k = 2, 3 \qquad (10.45)$$

where ϵ_{k1} depends on temperature only. Introducing (10.45) into (10.44) results in

$$\boxed{p^{kl} = \epsilon_{k1}(T)\exp\left[\frac{v_k}{R_1 T}\left(\frac{p^{kl}}{n^1} - p_0\right)\right], \quad k = 2, 3}\qquad (10.46)$$

This equation is difficult to use since the unknown quantity also appears in the exponent. In the approximation $v_k = 0$, $k = 2, 3$ we have

$$p^{kl} \approx \epsilon_{k1}, \quad k = 2, 3 \qquad (10.47)$$

which can be evaluated from (10.45). The result follows from (10.20) and (10.21)

$$\ln \epsilon_{k1}(T) = \frac{1}{R_1 T}[h_k(0) - h_1(0) - T(s_k(0) - \overset{\circ}{s}_1(0))]$$

$$+ \frac{c_k - c_{p,1}}{R_1 T}(T - T_0) - \frac{c_k - c_{p,1}}{R_1}\ln\frac{T}{T_0} + \ln p_0, \quad k = 2, 3$$

$$(10.48)$$

Next we eliminate the undetermined quantity $\epsilon_{k1}(T)$ by setting $T = T_0$ in the previous equation. The result is

$$\ln \epsilon_{k1}(T_0) = \frac{1}{R_1 T_0}[h_k(0) - h_1(0) - T_0(s_k(0) - \overset{\circ}{s}_1(0))] + \ln p_0, \quad k = 2, 3$$

$$(10.49)$$

Subtracting (10.49) from (10.48) gives

$$\ln \frac{\epsilon_{k1}(T)}{\epsilon_{k1}(T_0)} = -\frac{1}{R_1}[l_{k1}(0) + (c_k - c_{p,1})T_0]\left(\frac{1}{T} - \frac{1}{T_0}\right) - \frac{c_k - c_{p,1}}{R_1}\ln\frac{T}{T_0}, \quad k = 2, 3$$

(10.50)

where the identity $(h_1 - h_k) = l_{k1}$ has been used. In order to have a compact notation, we write in the approximation (10.47)

$$p^{k1}(T) = p^{k1}(T_0)\exp\left[-v_{k1}\ln\frac{T}{T_0} - \beta_{k1}\left(\frac{1}{T} - \frac{1}{T_0}\right)\right]$$

$$v_{k1} = \frac{c_k - c_{p,1}}{R_1}, \qquad \beta_{k1} = \frac{1}{R_1}[l_{k1}(0) + (c_k - c_{p,1})T_0], \qquad k = 2, 3$$

$$p^{21}(T_0) = 6.1070 \text{ hPa}, \qquad p^{31}(T_0) = 6.1064 \text{ hPa}, \qquad T_0 = 273.15 \text{ K}$$

(10.51)

If an even better approximation is desired we could replace $p^{k1}(T)$, $p^{k1}(T_0)$ in (10.51) by $\epsilon_{k1}(T)$, $\epsilon_{k1}(T_0)$ and substitute the result into (10.46). However, the dependency of the saturation vapor pressure on the molar fraction is seldom accounted for in meteorological practice. Equation (10.51) is an approximate integration of the Clausius–Clapeyron equation which is known as the *Magnus formula*. It is more convenient, instead of using (10.51), to use the empirical approximation to (10.51) as given by

$$p^{k1}(\vartheta) = p^{k1}(\vartheta = 0) \, 10^{d_{k1}\vartheta/(b_{k1}+\vartheta)}, \qquad k = 2, 3 \quad \text{with}$$

$$d_{21} = 7.4475, \qquad d_{31} = 9.5, \qquad b_{21} = 234.9\,°\text{C}, \qquad b_{31} = 265.5\,°\text{C}$$

(10.52)

where ϑ is the temperature given in degrees Celsius.

10.6 The equilibrium melting pressure

The affinity for the melting pressure is defined by

$$\begin{aligned}
a_{32}(p, T) &= \mu_3(p, T) - \mu_2(p, T) \\
&= \left[\mathring{\mu}_3(T) - \mathring{\mu}_1(T)\right] - \left[\mathring{\mu}_2(T) - \mathring{\mu}_1(T)\right] + (v_3 - v_2)(p - p_0) \\
&= R_1 T \ln\frac{\epsilon_{31}}{\epsilon_{21}} + (v_3 - v_2)(p - p_0)
\end{aligned}$$

(10.53)

where use has been made of (9.36) and (10.45). In the approximation $v_3 = v_2 = 0$ we find with (10.47),

$$a_{32}(T) \approx R_1 T \ln\frac{p^{31}(T)}{p^{21}(T)}$$

(10.54)

where T, in this case, is the temperature of the *triple point*, $T_{trip} = 273.16$ K. In case of equilibrium ($a_{32} = 0$) we find an expression for the melting pressure from (10.53) which reads

$$p^{32}(T) = p_0 - \frac{R_1 T}{v_3 - v_2} \ln \frac{\epsilon_{31}}{\epsilon_{21}} \approx p_0 - \frac{R_1 T}{v_3 - v_2} \ln \frac{p^{31}(T)}{p^{21}(T)} \qquad (10.55)$$

where use has been made of the approximation (10.47). Finally, in the approximation $v_3 = v_2 = 0$ we find for the equilibrium case from (10.54)

$$p^{31}(T) = p^{21}(T) \qquad (10.56)$$

This defines the *temperature of the triple point*, $T_{trip} \approx 273.16$ K. The melting pressure at $T = T_0$ is 1013.25 hPa while at the triple point $p^{32}(T_{trip}) = 6.1114$ hPa. This shows the steepness of the slope of the melting curve of about $-100\,000$ hPa per degree Kelvin.

10.7 Humidity measures related to chemical equilibrium

In Chapter 8 we have introduced various humidity measures which were not related to chemical equilibrium. Now we wish to define the same humidity measures with respect to chemical equilibrium, which are commonly referred to as *saturation values*.

Before proceeding with the discussion, we need to recall that we have treated the condensed phases as pure substances which do not contain foreign materials such as salts. If mixtures occur the chemical potentials will differ from the pure phase and subsequently the saturation vapor pressure will also differ. Moreover, the geometry of the surface separating the phases must be accounted for when the particles are sufficiently small. For example, if water droplets are smaller than 10^{-5} m the so-called *curvature effect* cannot be neglected. In order to guarantee defined conditions we assume that the condensed particles are sufficiently large so we may treat them as plane surfaces. Curvature and solution effects will be treated in a textbook on cloud physics.

10.7.1 Relative humidity

The *relative humidity* is defined by the relation

$$U = U^{21} = \frac{p^1}{p^{21}} 100\% \qquad (10.57)$$

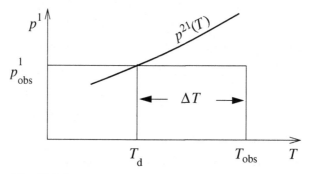

Fig. 10.2 Determination of the dew point temperature.

In case of saturation we have $a_{21} = 0$ and $U^{21} = 100\%$. It is customary to define the relative humidity with respect to p^{21} even for temperatures below 0 °C. If the air is supersaturated then the relative humidity exceeds 100%.

10.7.2 Dew point temperature

The *dew point temperature* T_d is that temperature that a parcel of air will obtain if it is cooled isobarically to saturation. During this process no moisture flux is permitted between the parcel of air and the surroundings. This means that we have to find a solution to the equation

$$\boxed{p^1 = p^{21}(T_d)} \tag{10.58}$$

The solution of this equation is best demonstrated by a graphical procedure as shown in Figure 10.2. In the (p^1, T)-plane we plot the observed point (p^1_{obs}, T_{obs}) and draw an isobar as shown in the figure. The intersection of the isobar p^1_{obs} with the saturation vapor pressure curve $p^{21}(T)$ defines the dew point temperature T_d. The quantity ΔT is the difference between the actually observed and the dew point temperature. If the saturation curve in the (p^1, T)-plane is p^{31} instead of p^{21} then one speaks of the *frost point temperature*.

Using the graphical method is very tedious. Therefore, we will introduce an analytic procedure using the empirical Magnus formula (10.52), which is written in terms of degrees Celsius. All we need to do is to replace the saturation vapor pressure $p^{21}(T)$ by the actual vapor pressure p^1 and the temperature by the dew point temperature in degrees Celsius which we will denote by the symbol τ. Instead of (10.52) we now write

$$p^1 = p^{21}(\vartheta = 0\,°C)\, 10^{d_{21}\tau/(b_{21}+\tau)}, \qquad \tau \quad \text{in} \quad °C \tag{10.59}$$

Solving for τ we obtain

$$\tau = -b_{21}\left(\frac{d_{21}}{\log_{10}[p^1/p^{21}(0)] - d_{21}} + 1\right) \qquad (10.60)$$

which is the desired relationship.

10.7.3 Dew point temperature difference

The *dew point temperature difference* is defined by

$$\Delta T = T - T_d = \vartheta - \tau = \Delta \tau$$

$$\Delta \tau(\vartheta, p^1) = \vartheta + b_{21} + \frac{b_{21}d_{21}}{\log_{10}[p^1/p^{21}(0)] - d_{21}} \qquad (10.61)$$

This difference, in contrast to τ itself, is a true humidity measure of the moist atmosphere since it depends on the vapor pressure p^1 and the temperature ϑ.

The meteorological practice requires a relationship between the relative humidity U and $\Delta \tau$. This relationship is easily found by observing that

$$\frac{p^1}{p^{21}(0)} = \frac{p^1}{p^{21}(\vartheta)}\frac{p^{21}(\vartheta)}{p^{21}(0)} = \frac{U}{100\%}\frac{p^{21}(\vartheta)}{p^{21}(0)} \qquad (10.62)$$

The ratio $p^{21}(\vartheta)/p^{21}(0)$ can be found from equation (10.52). Therefore,

$$\frac{p^1}{p^{21}(0)} = \frac{U}{100\%}10^{d_{21}\vartheta/(b_{21}+\vartheta)} \qquad (10.63)$$

Substituting $\log(p^1/p^{21})$ from this expression into (10.61) gives the desired relationship

$$\Delta \tau(\vartheta, U) = \frac{\log_{10}\left(\frac{U}{100\%}\right)(b_{21} + \vartheta)}{\log_{10}\left(\frac{U}{100\%}\right) - \frac{b_{21}d_{21}}{b_{21}+\vartheta}} \qquad (10.64)$$

from which the value of U can be found if $\Delta \tau$ is known.

10.7.4 Saturation mixing ratio and specific humidity

In Chapter 8, equations (8.84) and (8.85), we have introduced the mixing ratio r and the specific humidity q. Solving (8.87) and using (8.86) we obtain

$$r = \frac{R_0}{R_1}\frac{n^1}{1 - n^1} \approx \frac{R_0}{R_1}n^1, \qquad q = \frac{R_0}{R_1}\frac{n^1}{1 + n^1(R_0/R_1 - 1)} \approx \frac{R_0}{R_1}n^1 \qquad (10.65)$$

With the help of Dalton's law, i.e. $p^1 = n^1 p$, and assuming saturation conditions $p^1 = p^{21}$, we can write

$$
\begin{aligned}
r^{21}(p, T) &= \frac{R_0}{R_1} \frac{p^{21}(T)}{p - p^{21}(T)} \approx \frac{R_0}{R_1} \frac{p^{21}(T)}{p} \\
q^{21}(p, T) &= \frac{R_0}{R_1} \frac{p^{21}(T)}{p + p^{21}(T)(R_0/R_1 - 1)} \approx \frac{R_0}{R_1} \frac{p^{21}(T)}{p}
\end{aligned}
\tag{10.66}
$$

In the approximation $p^{21} \ll p$ there is no difference between the saturation mixing ratio r^{21} and the saturation specific humidity q^{21}. The important point is that in case of saturation when $a_{21} = 0$ the quantities r^{21} and q^{21} are no longer independent coordinates but they depend uniquely on p and T.

We now form the ratios r/r^{21} and q/q^{21} and find approximation formulas for the relative humidity

$$
\frac{q}{q^{21}} \approx \frac{r}{r^{21}} \approx \frac{p^1}{p^{21}} = \frac{U}{100\%} \implies U \approx \frac{q}{q^{21}} 100\%
\tag{10.67}
$$

which are sufficiently accurate for most practical purposes. In the case that we wish to apply r and q to saturation with respect to ice, we replace p^{21} by p^{31} and use the appropriate tables or formulas.

10.7.5 Virtual temperature correction

In equation (8.89) we replace n^1 by p^1/p according to Dalton's law and obtain in the case of saturated and unsaturated air

$$
\Delta T_v \approx 0.61 \frac{R_0 T}{R_1 p} p^1, \qquad \Delta T_v^{21} \approx 0.61 \frac{R_0 T}{R_1 p} p^{21}
\tag{10.68}
$$

Taking the ratio $\Delta T_v / \Delta T_v^{21}$, we obtain a useful expression for the relative humidity

$$
\frac{\Delta T_v}{\Delta T_v^{21}} = \frac{p^1}{p^{21}} = \frac{U}{100\%} \implies U = \frac{\Delta T_v}{\Delta T_v^{21}} 100\%
\tag{10.69}
$$

If needed, the saturation can be expressed with respect to ice.

10.7.6 Wet bulb temperature

At the ground the water vapor content of the moist air is usually measured with the help of an *Assmann ventilated psychrometer*. Essentially this instrument consists of the *dry bulb thermometer* and the *wet bulb thermometer*. The wet bulb thermometer

Wet bulb thermometer   *Dry bulb thermometer*

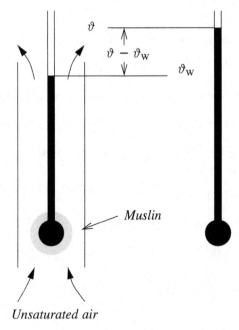

Fig. 10.3 The wet bulb process.

is in reality an ordinary thermometer whose bulb is covered with a wet piece of muslin. For temperatures below freezing the muslin will be iced. By means of a motor, a fan is caused to draw air past the thermometer bulbs. The heat required to evaporate the water or the ice in the muslin is taken from the air which in the stationary case is cooled from the dry bulb temperature ϑ to the wet bulb temperature ϑ_w which are expressed in degrees Celsius. The situation is depicted in Figure 10.3.

We simulate the thermodynamic process at the wet bulb by assuming a closed system which is taken from the initial state I (before evaporation) by an isobaric and adiabatic process: $dp = 0$, $đ(q + w) = 0$ to state II after evaporation. Starting with the first law in the form (3.36) we find $dh = v\,dp + đ(q + w) = 0$ so that

$$h_{\mathrm{I}} = h_{\mathrm{II}} \quad \text{or} \quad \text{extensively} \quad H_{\mathrm{I}} = H_{\mathrm{II}} \tag{10.70}$$

indicating that the enthalpy is being conserved. Since the system is assumed to be closed, $d_e M^k = 0$, $k = 0, \ldots, 3$, we have $dM^k = d_i M^k$, $dM^0 = 0$, $dM^{H_2O} = 0$. This is equivalent to

$$M_{\mathrm{I}}^0 = M_{\mathrm{II}}^0 = M^0, \qquad M_{\mathrm{I}}^{H_2O} = M_{\mathrm{II}}^{H_2O} \tag{10.71}$$

We assume that the atmospheric state II refers to saturation so that

$$p_{\mathrm{II}}^1 = p^{21}(T), \qquad r_{\mathrm{II}} = r^{21}(p, T) \tag{10.72}$$

In case that the water in the muslin is frozen, the superscript ^{21} denoting equilibrium between water and water vapor must be replaced by ^{31} referring to equilibrium between ice and water vapor.

From (1.16) we find $H = h_n M^n$ so that with $H_{\mathrm{I}} = H_{\mathrm{II}}$ we obtain

$$M_{\mathrm{I}}^0 h_0(\vartheta) + M_{\mathrm{I}}^1 h_1(\vartheta) + M_{\mathrm{I}}^2 h_2(p, \vartheta_{\mathrm{w}}) = M_{\mathrm{II}}^0 h_0(\vartheta_{\mathrm{w}}) + M_{\mathrm{II}}^1 h_1(\vartheta_{\mathrm{w}}) \tag{10.73}$$

Dividing this expression by $M_{\mathrm{I}}^0 = M_{\mathrm{II}}^0$ and introducing the mixing ratio we find with $M_{\mathrm{I}}^2 = M_{\mathrm{II}}^1 - M_{\mathrm{I}}^1$ the expression

$$h_0(\vartheta) - h_0(\vartheta_{\mathrm{w}}) + rh_1(\vartheta) + [r^{21}(p, \vartheta_{\mathrm{w}}) - r]h_2(p, \vartheta_{\mathrm{w}}) = r^{21}(p, \vartheta_{\mathrm{w}})h_1(\vartheta_{\mathrm{w}}) \tag{10.74}$$

With $h_1 - h_2 = l_{21}$ this expression can be rewritten as

$$h_0(\vartheta) - h_0(\vartheta_{\mathrm{w}}) + rh_1(\vartheta) + [r^{21}(p, \vartheta_{\mathrm{w}}) - r][h_1(\vartheta_{\mathrm{w}}) - l_{21}(p, \vartheta_{\mathrm{w}})] = r^{21}(p, \vartheta_{\mathrm{w}})h_1(\vartheta_{\mathrm{w}}) \tag{10.75}$$

Using equation (8.83) we readily find

$$c_{p,0}(\vartheta - \vartheta_{\mathrm{w}}) + rc_{p,1}(\vartheta - \vartheta_{\mathrm{w}}) = [r^{21}(p, \vartheta_{\mathrm{w}}) - r]l_{21}(p, \vartheta_{\mathrm{w}}) \tag{10.76}$$

With the approximation (10.66) for the saturation mixing ratio we obtain

$$\vartheta - \vartheta_{\mathrm{w}} = \frac{\left(\frac{R_0}{R_1} \frac{p^{21}(\vartheta_{\mathrm{w}})}{p} - r \right) l_{21}(p, \vartheta_{\mathrm{w}})}{c_{p,0} + rc_{p,1}} \tag{10.77}$$

For practical purposes we introduce the quite acceptable approximations $c_{p,0} \gg rc_{p,1}$, $l_{21}(\vartheta_{\mathrm{w}}, p) \approx l_{21}(\vartheta = 0)$ and find

$$\vartheta - \vartheta_{\mathrm{w}} = \frac{R_0 l_{21}(0)}{R_1 p c_{p,0}} [p^{21}(\vartheta_{\mathrm{w}}) - p^1], \quad \text{or}$$

$$p^1 = p^{21}(\vartheta_{\mathrm{w}}) - \frac{p(\vartheta - \vartheta_{\mathrm{w}})}{C_{21}}, \quad \text{with} \quad C_{21} = \frac{R_0 l_{21}(0)}{R_1 c_{p,0}} = 1550\,\mathrm{K} \tag{10.78}$$

The term C_{21} is known as *Sprung's constant*.

With p^1 and $p^{21}(T)$ known, we easily find the relative humidity U since $p^{21}(T)$ can be calculated or extracted from tables. In case of frozen water in the muslin we must use different values of Sprung's constant. Instead of p^{21} and l_{21} we use p^{31} and l_{31} we find as a new value for Sprung's constant

$$C_{31} = \frac{C_{21}}{0.89} \tag{10.79}$$

In practice one uses the so-called *psychrometer tables* either for water or ice. One enters the tables with the measured values of p, ϑ and ϑ_{w} and finds p^1, U and τ.

10.8 Appendix

In this Appendix some additional equations on the evaporation affinity a_{k1} will be presented. Combination of (10.42) with (9.36) results in

$$a_{k1}(p, T, n^1) = \overset{+}{\mu}_k(T) - \overset{+}{\mu}_1(T) + \left(\frac{p^1}{n^1} - p_0\right)v_k - R_1 T \ln p^1 \qquad (10.80)$$

Introducing (10.45) yields

$$a_{k1}(p, T, n^1) = R_1 T \ln \epsilon_{k1} + \left(\frac{p^1}{n^1} - p_0\right)v_k - R_1 T \ln p^1 \qquad (10.81)$$

In case of chemical equilibrium $a_{k1} = 0$ and $p^1 = p^{k1}$ so that (10.81) reduces to

$$0 = R_1 T \ln \epsilon_{k1} + \left(\frac{p^{k1}}{n^1} - p_0\right)v_k - R_1 T \ln p^{k1} \qquad (10.82)$$

Subtracting (10.82) from (10.81) gives the desired result

$$a_{k1}(p^1, T, n^1) = R_1 T \ln \left(\frac{p^{k1}}{p^1}\right) + (p^1 - p^{k1})\frac{v_k}{n^1}, \qquad k = 2, 3 \qquad (10.83)$$

Since $p^1 = n^1 p$ we may write $a_{k1}(p^1, T, n^1) = a_{k1}(p, T, n^1)$. For many practical applications it is permissible to ignore the second term on the right-hand side term of (10.83) in the approximation $v_k = 0$. Now the coefficients a_{k1} appearing in (7.88) can be evaluated. There exist extensive tables on the saturation vapor pressure.

10.9 Problems

10.1: At normal pressure 50 g of ice are evaporated by increasing the temperature from $\vartheta = -40\,°C$ to $\vartheta = 100\,°C$.
Calculate (a) the amount of heat to be added and (b) the entropy change. The following constants are given: $c_2 = 4200\,J\,kg^{-1}\,K^{-1}$, $c_3 = 2100\,J\,kg^{-1}\,K^{-1}$, $l_{21} = 2260 \times 10^3\,J\,kg^{-1}$, $l_{31} = 336 \times 10^3\,J\,kg^{-1}$.

10.2: Determine (a) the specific volume v_{cr}, (b) the temperature T_{cr}, and (c) the pressure p_{cr} at the critical point of a van der Waals gas defined by

$$p = \frac{R^* T}{\check{v} - b} - \frac{a}{\check{v}^2}$$

The quantities a, b are constants. Above their critical temperature gases cannot be liquified even by application of very high pressures. The critical isotherm has no extreme values but an inflexion point defined by the conditions

$$\left(\frac{\partial p}{\partial \check{v}}\right)_{T_{cr}} = 0, \qquad \left(\frac{\partial^2 p}{\partial \check{v}^2}\right)_{T_{cr}} = 0$$

10.3: Show that the finite change of enthalpy between the states (p_i, T_i) and (p_f, T_f) can be written as

$$h_f - h_i = \int_i^f c_p dT + \int_i^f v(1 - v^* T) dp$$

where i and f denote the initial and final state.

10.4: Assume that water vapor in equilibrium with a horizontal ice surface can be treated as an ideal gas. Calculate the heat of sublimation at $-20\,°C$ in $J\,kg^{-1}$ from the following experimental data:

ϑ (°C)	p^{31} (hPa)
-19.5	1.077
-20.0	1.067
-20.5	0.979

10.5: Prove that the slope of the sublimation curve of ice at the triple point is larger than the evaporation curve at the same point.

10.6: The boiling point of water decreases with increasing height above sea-level.

(a) At what temperature does water boil on top of mount Everest (8900 m) where the air pressure is 325 hPa?
(b) At what height does the human blood boil in an unpressurized cabin? The blood temperature is 37 °C. Assume that $l_{21} = 2226 \times 10^3\ J\,kg^{-1}$.

10.7: A water cloud of initial temperature T_i and initial water content m_i^2 evaporates. Find the temperature T_f of the air if all the liquid water evaporates isobarically and without change of the enthalpy. Assume a closed system and that l_{21} is independent of pressure.

10.8: Find the change in the melting temperature ΔT_{melt} of ice if the pressure is increased by 10^5 hPa.

11

Heat equation and special adiabatic systems

The heat equation or the prognostic equation for atmospheric temperature has been derived in detail in Chapter 3. The numerical evaluation of this equation is extremely laborious and very difficult. For practical purposes it is often sufficient to evaluate approximate forms of this equation. Some of these will be derived and discussed in the following sections. Of particular interest are the adiabatic approximations which are used later to construct thermodynamic diagrams that are indispensable in the daily routines of any weather service.

11.1 The modified heat equation

First of all, we will modify the exact form of the heat equation by introducing slight approximations such as ignoring the specific volumes of water and ice particles. According to (10.12b) and (10.13), the specific heat of tension γ and the gradient of the enthalpy ∇h_k can then be replaced by $-v$ and $c_{p,k}\nabla T$, respectively. Equation (3.52) then reduces to

$$\rho c_p \frac{dT}{dt} = \frac{dp}{dt} + l_{21}I^2 + l_{31}I^3 - \nabla \cdot \left(\mathbf{J}_s^h + \mathbf{F_R}\right) - c_{p_n}\mathbf{J}^n \cdot \nabla T + \mathbb{J} \cdots \nabla \mathbf{v} \quad (11.1)$$

This equation in conjunction with the prognostic equations for the barycentric velocity and the concentrations, together with the continuity equation for the total density, constitute a complete prognostic system. The diagnostic equations to be used are the following:

(i) the phenomenological equations for the various fluxes,
(ii) the equation of state,
(iii) the radiative transfer equation.

165

11.2 The adiabatic heat equation

Whenever short-term atmospheric developments are considered, with sufficient degree of approximation, we may often assume that no heat due to fluxes of sensible heat, radiation and energy dissipation is added to the system. The system is not closed with respect to mass so that the diffusion flux will not vanish. The conditions that regulate the *adiabatic process* are then given by

$$\nabla \cdot \left(\mathbf{J}_s^h + \mathbf{F}_R \right) = 0, \qquad \mathbb{J} \cdot\cdot \nabla \mathbf{v} = 0 \tag{11.2}$$

The adiabatic heat equation then reads

$$\rho c_p \frac{dT}{dt} = \frac{dp}{dt} + l_{21} I^2 + l_{31} I^3 - c_{p_n} \mathbf{J}^n \cdot \nabla T \tag{11.3}$$

An equivalent equation can also be written for the coordinates T and v.

11.3 Thermodynamic filtering of the heat equation

In order to avoid problems with the evaluation of the phase transition fluxes I^k, one often assumes *chemical equilibrium* between the water vapor and one of the condensed phases. Thus the actual adjustment process to the chemical equilibrium is eliminated or filtered. In fact, phase transition fluxes and chemical affinities do not even have to be considered since the mixing ratio of the cloud air depends uniquely on the state coordinates p and T. The price to be paid is that the continuity equations for the concentrations will have a more complicated mathematical structure.

It should be pointed out that the filtering process is only possible between water vapor and one of the condensed phases since all three phases exist only at the triple point where temperature and pressure are fixed quantities. In practical cloud modeling this would require a fixed isotherm such as $T = 253$ K to separate the water phase from the ice. Particles transported across this boundary must then change phase immediately. Due attention must be paid to latent heat effects.

The filtering process is then expressed by the condition

$$a_{k1} = 0 \implies p^1 = p^{k1}(T), \qquad r = \frac{m^1}{m^0} = r^{k1}(p, T) \tag{11.4}$$

with $m^3 = 0$ for a water cloud ($k = 2$) and $m^2 = 0$ for an ice cloud ($k = 3$). An immediate consequence is that the prognostic equations for the concentrations and the heat equation must be modified. We will now turn to this task.

11.3.1 Filtering of the prognostic equations for the concentrations

The prognostic equation for the concentration m^0 is not affected by the filtering process. We introduce the filter condition into the prognostic equation for m^1.

Therefore, we can write

$$\rho \frac{dm^0}{dt} = -\nabla \cdot \mathbf{J}^0 \quad \text{since} \quad I^0 = 0$$

$$\rho \frac{dm^1}{dt} = \rho \frac{d}{dt}(m^0 r^{k1}) = r^{k1} \rho \frac{dm^0}{dt} + \rho m^0 \frac{dr^{k1}}{dt} \qquad (11.5)$$

$$= -r^{k1} \nabla \cdot \mathbf{J}^0 + \rho m^0 \frac{dr^{k1}}{dt}$$

What we need is a relationship between the *saturation mixing ratio* and the *saturation vapor pressure*. Solving the defining equation (10.66) for the saturation vapor pressure and taking logarithms, we obtain the following differential expression

$$d \ln p^{k1} = d \ln p + \frac{dr^{k1}}{r^{k1}(1 + r^{k1} R_1 / R_0)} \qquad (11.6)$$

Now substituting for $d \ln p^{k1}$ from the Clausius–Clapeyron equation (10.43b) with $p^1(a_{k1} = 0) = p^{k1}$ we find the relationship

$$dr^{k1} = \left(\frac{l_{k1}}{R_1 T^2} dT - \frac{dp}{p} \right) \left(1 + r^{k1} \frac{R_1}{R_0} \right) r^{k1} \qquad (11.7)$$

For many practical purposes we may neglect $r^{k1} R_1 / R_0$ in comparison to 1. Substituting (11.7) into (11.5) gives the desired set of filtered equations. The system to be solved is then given by

$$\boxed{\begin{aligned} \rho \frac{dm^0}{dt} &= -\nabla \cdot \mathbf{J}^0 \\ \rho \frac{dm^1}{dt} &= -r^{k1} \nabla \cdot \mathbf{J}^0 + \rho m^0 \left(\frac{l_{k1}}{R_1 T^2} \frac{dT}{dt} - \frac{1}{p} \frac{dp}{dt} \right) \left(1 + r^{k1} \frac{R_1}{R_0} \right) r^{k1} \\ m^k &= 1 - m^0 - m^1, \qquad k = 2, 3 \end{aligned}} \qquad (11.8)$$

Only two prognostic equations need to be evaluated since the water vapor can be in equilibrium only with one of the condensed phases. The third equation for m^k, $k = 2, 3$ can be found diagnostically as stated in (11.8). In a later chapter we will demonstrate the filtering procedure by using a different method.

11.3.2 Filtering of the prognostic equation of temperature

Since only one of the condensed phases is participating, we may write instead of (11.1) the following expression

$$\rho c_p \frac{dT}{dt} = \frac{dp}{dt} + l_{k1} I^k - \nabla \cdot \left(\mathbf{J}_s^h + \mathbf{F}_R \right) - c_{p_n} \mathbf{J}^n \cdot \nabla T + \mathbb{J} \cdots \nabla \mathbf{v} \quad \text{with}$$

$$c_{p_n} \mathbf{J}^n \cdot \nabla T = \left(c_{p_0} \mathbf{J}^0 + c_{p_1} \mathbf{J}^1 + c_k \mathbf{J}^k \right) \cdot \nabla T, \qquad k = 2, 3$$

(11.9)

We will now replace the phase transition flux in (11.9) by

$$v I^k = \frac{d_i m^k}{dt} = -\frac{d_i m^1}{dt} = -\frac{dm^1}{dt} + \frac{d_e m^1}{dt}$$

$$= v r^{k1} \nabla \cdot \mathbf{J}^0 - m^0 \frac{d r^{k1}}{dt} - v \nabla \cdot \mathbf{J}^1$$

(11.10)

where use has been made of (2.20) and (11.5). Finally, substituting (11.10) into (11.9) gives the desired heat equation or the prognostic equation for the temperature in the filtered form

$$\boxed{ \begin{aligned} c_p \frac{dT}{dt} = {} & v \frac{dp}{dt} + v l_{k1} r^{k1} \nabla \cdot \mathbf{J}^0 - m^0 l_{k1} \frac{d r^{k1}}{dt} - v l_{k1} \nabla \cdot \mathbf{J}^1 \\ & - v \nabla \cdot \left(\mathbf{J}_s^h + \mathbf{F}_R \right) - v c_{p_n} \mathbf{J}^n \cdot \nabla T + v \mathbb{J} \cdots \nabla \mathbf{v} \end{aligned} }$$

(11.11)

The expression $d\, r^{k1}/dt$ appearing in (11.11) is given by (11.7).

11.4 The heat equation of homogeneous systems

The starting point of the analysis is the heat equation (11.1) which is quite general and refers to an inhomogeneous system. To make this equation homogeneous, we ignore all gradient terms of the thermodynamic variables p, T, m^k. The gradients $\nabla p, \nabla T, \nabla m^k$ also appear implicitly in the individual derivatives so that these reduce to the local derivatives $d(\cdots)/dt \longrightarrow \partial(\cdots)/\partial t$. Only one term containing a gradient (∇T) appears explicitly in (11.1). Therefore, we may write

$$c_p \frac{dT}{dt} = v \frac{dp}{dt} + v l_{21} I^2 + v l_{31} I^3 - v \nabla \cdot \mathbf{J}_s^h + v \mathbb{J} \cdots \nabla \mathbf{v}$$

(11.12)

We have also omitted the radiative flux divergence term since it has a tendency to destroy homogeneous conditions.

In order to convert (11.12) to the extensive form it is best to first replace all fluxes as shown next.

$$\frac{d_i m^k}{dt} = v I^k, \qquad \frac{d_e m^k}{dt} = -v \nabla \cdot \mathbf{J}^k, \qquad \frac{d_s q}{dt} = -v \nabla \cdot \mathbf{J}_s^h, \qquad \frac{d w}{dt} = v \mathbb{J} \cdots \nabla \mathbf{v}$$

(11.13)

To indicate that the flux divergence of the sensible heat is meant we have also added the subscript $_s$ to the time derivative. Substituting (11.13) into (11.12) and then multiplying the resulting equation by dt gives

$$c_p dT = v\,dp + l_{21}d_i m^2 + l_{31}d_i m^3 + d_s q + d w \tag{11.14}$$

This equation could have been derived also by starting with the first law of thermodynamics in the form (3.36) which is valid for open as well as for closed systems. Next we multiply this equation by the total mass M of the system, which is not necessarily constant. Using the definitions

$$C_p = c_p M, \qquad d_i M = 0, \qquad d_e M \neq 0, \qquad M d_s q = d_s Q, \qquad M d w = d W \tag{11.15}$$

we obtain

$$C_p dT = V\,dp + l_{21}d_i M^2 + l_{31}d_i M^3 + d_s Q + d W \tag{11.16}$$

This equation is valid for open as well as for closed systems. The reason that mass transport terms do not occur is easily explained. The physical coordinates p and T cannot change by an exchange of mass having the uniform temperature and pressure, which characterize the homogeneous system.

This section will be concluded by showing the difference between $M d_s q = d_s Q$ and $M d q = d Q$. Using the definition (3.51) we find

$$\frac{dq}{dt} = -v \nabla \cdot \mathbf{J}^h = -v \nabla \cdot \left(\mathbf{J}_s^h + h_n \mathbf{J}^n \right) \tag{11.17}$$

and recalling that for a homogeneous system $\nabla h_k = 0$, we have

$$\frac{dq}{dt} = \frac{d_s q}{dt} + h_n \frac{d_e m^n}{dt}$$
$$\frac{dQ}{dt} = \frac{d_s Q}{dt} + h_n \frac{d_e M^n}{dt} - h \frac{d_e M}{dt} \tag{11.18}$$
$$= \frac{d_s Q}{dt} + \sum_{k=0}^{3}(h_k - h)\frac{d_e M^k}{dt}$$

where the extensive form was obtained by multiplying the intensive formulation by the total mass M.

11.5 The adiabatic heat equation of homogeneous systems

According to (11.2) the adiabatic process is defined by

$$\frac{d_s q}{dt} = -v \nabla \cdot \mathbf{J}_s^h = 0 \quad \text{or extensively} \quad \frac{d_s Q}{dt} = 0$$
$$\frac{d w}{dt} = v \mathbf{J} \cdots \nabla \mathbf{v} = 0 \quad \text{or extensively} \quad \frac{d W}{dt} = 0 \tag{11.19}$$

where the radiative flux was omitted already. Applying (11.19) to (11.14) and (11.16) gives the desired result

$$\boxed{C_p dT = V dp + l_{21} d_i\, M^2 + l_{31} d_i\, M^3}$$
(11.20)

or intensively

$$\boxed{c_p dT = v dp + l_{21} d_i m^2 + l_{31} d_i m^3}$$
(11.21)

This equation is the starting point for further discussions as presented in the next chapter.

11.6 Problems

11.1: Start with the first law of thermodynamics in the form $dh - v dp = đq + đw$ and derive equation (11.16).

11.2: Show that for a closed filtered system the continuity equations for the partial masses can be written as

$$\frac{dm^0}{dt} = 0, \qquad \frac{dm^k}{dt} = -m^0 \left(\frac{\partial r^{k1}}{\partial T}\frac{dT}{dt} + \frac{\partial r^{k1}}{\partial p}\frac{dp}{dt} \right)$$

11.3: Show that the filtered heat equation can be written in the form

$$\rho c_p \left(1 + \frac{m^0 l_{k1}^2 \overset{*}{r}{}^{k1}}{c_p R_1 T^2} \right) \frac{dT}{dt} = \left(1 + \frac{m^0 \rho l_{k1} \overset{*}{r}{}^{k1}}{p} \right) \frac{dp}{dt} + B_{k1}$$

$$\text{with} \quad \overset{*}{r}{}^{k1} = r^{k1}\left(1 + r^{k1}\frac{R_1}{R_0} \right)$$

$$B_{k1} = l_{k1}(r^{k1}\nabla\cdot\mathbf{J}^0 - \nabla\cdot\mathbf{J}^1) - \nabla\cdot\mathbf{J}_s^h + \mathbb{J}\cdots\nabla\mathbf{v} - \mathbf{J}^n c_{p,n}\cdot\nabla T$$

12

Special adiabats of homogeneous systems

In order to describe adiabatic processes in homogeneous thermodynamic systems we make use of (11.21). This equation is valid for open as well as for closed systems. For a homogeneous closed system the state curve due to adiabatic processes is called an *adiabat*. If the system is open the state curve is called a *pseudoadiabat*. In this chapter we will consider both situations and give precise mathematical derivations.

12.1 Adiabats of an unsaturated system

The most simple *adiabatic process* we can think of refers to a closed homogeneous system of unsaturated air. The task ahead is to find a relationship between temperature and pressure for such a process. Application of (11.21) with the requirement

$$m^2 = m^3 = 0, \qquad d_i m^1 = 0, \qquad d_e m^0 = d_e m^1 = 0 \tag{12.1}$$

results in

$$\left(c_{p,0} m^0 + c_{p,1} m^1\right) \frac{dT}{T} - \left(R_0 m^0 + R_1 m^1\right) \frac{dp}{p} = 0 \tag{12.2}$$

The coefficient k_{m} is a constant of the system as given by

$$\frac{dT}{T} - k_{\mathrm{m}} \frac{dp}{p} = 0 \quad \text{with} \quad k_{\mathrm{m}} = \frac{R_0 m^0 + R_1 m^1}{c_{p,0} m^0 + c_{p,1} m^1} = \text{constant} \tag{12.3}$$

In case of dry air ($m^1 = 0$, $m^0 = 1$) we have $k_{\mathrm{m}} = R_0/c_{p,0} = k_0$ so that

$$k_{\mathrm{m}} = k_0 \frac{1 + r R_1/R_0}{1 + r c_{p,1}/c_{p,0}} \tag{12.4}$$

The coefficients k_0 and k_{m} are called the *exponents of the adiabats for dry and moist air*, respectively.

In order to find the required (p, T)-relationship of the adiabatic process, we integrate (12.3) with the result that

$$T = \text{constant } p^{k_m} \tag{12.5a}$$

which is an *adiabat of the moist air*. To evaluate the integration constant it is conventional to introduce the *potential temperature of the moist air* θ_m. Whenever the system assumes the fixed pressure $p = p_0 = 1000$ hPa, the temperature T is replaced by θ_m so that

$$\theta_m = \text{constant } p_0^{k_m} \tag{12.5b}$$

Elimination of the constant then results in

$$\boxed{\theta_m = T \left(\frac{p_0}{p} \right)^{k_m}} \tag{12.6}$$

This is the general definition of the potential temperature of moist air. From (12.5b) follows that $d\theta_m = 0$ so that θ_m is a constant of an individual system or a conservative quantity for this particular process. In general, however, as follows from (12.6), the moist potential temperature can be expressed in the form $\theta_m = \theta_m(p, T, k_m) = \theta_m(p, T, r)$. The exponent k_m is not a real constant but depends on the mixing ratio r. Through every point in the (p, T)-plane we may draw a family of adiabats which vary with r. For an individual system r is fixed, in general r varies from 0 to 1. For most practical purposes we may replace k_m by k_0 since $r \ll 1$. From (12.4) we see that $k_m = k_0$ whenever $r = 0$. Therefore, the potential temperature of dry air is defined by

$$\boxed{\theta = T \left(\frac{p_0}{p} \right)^{k_0}} \tag{12.7}$$

If an air parcel is displaced dry adiabatically, then θ is a constant of the system so that $d\theta/dt = 0$.

In conclusion we remark that dry adiabats $\theta = $ constant play a prominent role in thermodynamic charts, see Chapter 13, which are used to find graphical solutions of thermodynamic problems.

12.2 The equivalent temperature

Consider a closed homogeneous system consisting of dry air, water vapor and liquid water or ice. We will now proceed to find a relationship between the temperature T, the water vapor concentration m^1 and m^k, $k = 2, 3$ if the system is restricted to move adiabatically and isobarically. The system then is required to obey the following conditions:

$d_e m^k = 0$ for all k (closed system), $dm^k = d_i m^k$, $m^0 + m^1 + m^k = $ constant. Since $d_i m^0 = 0$ we have $dm^0 = 0$ and, therefore, $d_i m^1 + d_i m^k = 0, d_s q = 0,$ $dw = 0$ (adiabatic process), $dp = 0$ (isobaric process).

Application of these conditions to (11.21) results in

$$(c_{p,0} m^0 + c_{p,1} m^1 + c_k m^k) dT = l_{k1} d_i m^k = -l_{k1} d_i m^1 = -l_{k1} dm^1 \qquad (12.8)$$

Using the identity $l_{k1} dm^1 = d(l_{k1} m^1) - m^1 dl_{k1}$ we find with the help of (10.39b) and $v_k = 0, \ k = 2, 3$

$$(c_{p,0} m^0 + c_{p,1} m^1 + c_k m^k) dT + m^1 (c_k - c_{p,1}) dT + d(l_{k1} m^1) = 0 \qquad (12.9)$$

Abbreviating $m^1 + m^k = m^{H_2O}$ we obtain

$$\left(c_{p,0} m^0 + c_k m^{H_2O}\right) dT + d(l_{k1} m^1) = 0 \qquad (12.10)$$

Using the definition $r^{H_2O} = m^{H_2O}/m^0$ and recalling that m^0 is a constant, we may finally write

$$\left(c_{p,0} + c_k r^{H_2O}\right) dT + d(l_{k1} r) = 0 \qquad (12.11)$$

Integration of this equation gives immediately

$$T + \frac{l_{k1} r}{c_{p,0} + c_k r^{H_2O}} = C \qquad (12.12)$$

The constant C refers to a special temperature which will be defined next. For this purpose we visualize an idealized process causing all water vapor in the system to condense (water cloud) or to sublimate (ice cloud) completely so that at the end of the process $r = 0$. The special temperature thus obtained is the *equivalent temperature* $C = T_e$ or

$$\boxed{T_e = T + \frac{l_{k1} r}{c_{p,0} + c_k r^{H_2O}}, \qquad k = 2, 3} \qquad (12.13)$$

While r and T are state variables, r^{H_2O} and T_e are constants of the system, i.e. conservative quantities.

In practical meteorology the equivalent temperature usually refers to moist air consisting of dry air and water vapor, i.e. $r^{H_2O} = r$ so that

$$(T_e)_m \approx T + \frac{l_{k1} r}{c_{p,0}} \qquad (12.14)$$

since for realistic situations $c_{p,0} \gg r c_k$. T_e is related to the enthalpy of the system which can be used to measure the degree of human comfort since it involves the temperature and the humidity of the air. High values of T and r represent the maritime tropical air mass.

12.3 The reversible moist adiabat

In order to generalize the concept of the dry adiabat, we consider a homogeneous closed system consisting of dry air, water vapor and liquid water. The same formulation also applies to an ice cloud if m^2 is replaced by m^3 with corresponding changes of the other pertinent variables.

The physical system to be described requires that the phase transition between water vapor and liquid water takes place reversibly, i.e. chemical equilibrium between the two phases is required. This is the so-called *filtered system*. Now we are going to derive a differential relationship between the pressure and the temperature of the system that is known as the reversible moist adiabat. The assumptions leading to the reversible moist adiabat are summarized next.

$$d_e m^k = 0, \qquad dm^k = d_i m^k, \qquad m^0 = \text{constant}$$
$$m^1 + m^2 = \text{constant}, \qquad m^3 = 0, \qquad a_{21} = 0 \tag{12.15}$$

These conditions imply that the system is moving along the saturation vapor pressure curve p^{21} in the (p^1, T)-plane.

We start the analysis by dividing (11.21) by T and setting $m^3 = 0$ so that

$$c_p \frac{dT}{T} - \frac{v}{T} dp = \frac{l_{21}}{T} d_i m^2 = -\frac{l_{21}}{T} d_i m^1 = -\frac{l_{21}}{T} dm^1 \quad \text{with}$$

$$c_p = c_{p,0} m^0 + c_{p,1} m^1 + c_2 m^2, \qquad \frac{v}{T} = \frac{R_0 m^0 + R_1 m^1}{p}, \qquad v_2 = v_3 = 0 \tag{12.16}$$

Next we rewrite the right-hand side of (12.16) by using the approximation $dl_{21} = -(c_2 - c_{p,1}) dT$ which follows from (10.39b) by setting $v_k = 0, \ k = 2, 3$. This gives immediately

$$\frac{l_{21}}{T} dm^1 = \frac{d(l_{21} m^1)}{T} + m^1 (c_2 - c_{p,1}) \frac{dT}{T}$$
$$= d\left(\frac{l_{21} m^1}{T}\right) + l_{21} m^1 \frac{dT}{T^2} + m^1 (c_2 - c_{p,1}) \frac{dT}{T} \tag{12.17}$$

Substitution of this expression into (12.16) yields

$$\left(c_{p,0} m^0 + c_2 m^{H_2O}\right) \frac{dT}{T} - \left(R_0 m^0 + R_1 m^1\right) \frac{dp}{p} + d\left(\frac{l_{21} m^1}{T}\right) + l_{21} m^1 \frac{dT}{T^2} = 0 \tag{12.18}$$

Now we divide (12.18) by the factor multiplying dT/T which leads to the introduction of the constant k_{21} characterizing the system,

$$k_{21} = \frac{R_0 m^0}{c_{p,0} m^0 + c_2 m^{H_2O}} = \frac{R_0}{c_{p,0} \left(1 + c_2 m^{H_2O}/c_{p,0} m^0\right)} \approx \frac{R_0}{c_{p,0}} = k_0 \tag{12.19}$$

Introducing (12.19) into (12.18) with $r = r^{21}(p, T) = m^1/m^0$ gives

$$\frac{dT}{T} - k_{21}\frac{dp}{p} - \frac{R_1}{R_0}r^{21}k_{21}\frac{dp}{p} + d\left(\frac{l_{21}k_{21}r^{21}}{R_0 T}\right) + \frac{l_{21}k_{21}r^{21}}{R_0 T^2}dT = 0 \qquad (12.20)$$

We will now manipulate this equation so that only complete differentials appear which make an analytic integration possible. The details of the manipulation do not teach any new thermodynamics, so we leave them to the reader. First of all, we use the Clausius–Clapeyron equation (10.43b) with $p^1 = p^{21}$ and solve for p^{21} from (10.66)

$$\frac{l_{21}}{R_1 T^2}dT = d\ln p^{21}, \qquad \frac{p^{21}}{p} = \frac{r^{21}}{R_0/R_1 + r^{21}} \qquad (12.21)$$

From

$$\ln\left(\frac{p^{21}}{p}\right) = \ln r^{21} - \ln\left(\frac{R_0}{R_1} + r^{21}\right), \qquad r^{21} = \frac{R_0 p^{21}}{R_1 p^0}, \qquad p^0 = p - p^{21} \qquad (12.22)$$

we find

$$\ln\left(\frac{R_0/R_1 + r^{21}}{R_0/R_1}\right) = -\ln\left(\frac{p^0}{p}\right) = -\ln\left(1 - \frac{p^{21}}{p}\right) \qquad (12.23)$$

and, therefore,

$$d\ln\left(\frac{R_0}{R_1} + r^{21}\right) = -d\ln\left(\frac{p^0}{p}\right) = -d\ln\left(1 - \frac{p^{21}}{p}\right) \qquad (12.24)$$

With the help of (12.24) we obtain two additional versions of the Clausius–Clapeyron equation which will be used later.

(a) $\quad \dfrac{l_{21}}{R_1 T^2}dT = d\ln p^{21} = d\ln p + \dfrac{R_0}{R_1 r^{21}}d\ln\left(\dfrac{R_0}{R_1} + r^{21}\right)$

(b) $\quad \dfrac{l_{21}}{R_1 T^2}dT = d\ln p - \dfrac{R_0}{R_1 r^{21}}d\ln\left(\dfrac{p^0}{p}\right) \qquad\qquad (12.25)$

(c) $\quad \dfrac{l_{21}}{R_1 T^2}dT = d\ln p - \dfrac{R_0}{R_1 r^{21}}d\ln\left(1 - \dfrac{p^{21}}{p}\right)$

Substitution of (12.25a) into (12.20) gives the required expression of the reversible moist adiabat or

$$\boxed{\frac{dT}{T} - k_{21}\frac{dp}{p} + d\left(\frac{l_{21}k_{21}r^{21}}{R_0 T}\right) + k_{21}d\ln\left(\frac{R_0}{R_1} + r^{21}\right) = 0} \qquad (12.26)$$

Table 12.1. *Saturation vapor pressure p^{21} and r^{21}/T at different temperatures*

T (K)	p^{21} (hPa)	r^{21}/T at 1000 hPa	r^{21}/T at 100 hPa
223	0.0635	1.773×10^{-7}	1.773×10^{-6}
273	6.1070	1.393×10^{-5}	1.393×10^{-4}
323	123.39	2.378×10^{-4}	2.378×10^{-3}

In this form the reversible moist adiabat can be integrated immediately as shown next

$$\ln T - \ln p^{k_{21}} + \ln p_0^{k_{21}} + \frac{l_{21} k_{21} r^{21}}{R_0 T} + k_{21} \ln \left(\frac{R_0}{R_1} + r^{21} \right) = \ln C \qquad (12.27)$$

The constant $\ln(p_0^{k_{21}})$ could have been included in the integration constant. This equation leads to the definition of the potential equivalent temperature. The idea behind this definition is somewhat complex and will be discussed next.

First of all consider the idealized process as stated by the following limit

$$\lim_{T \to 0} \left[T \left(\frac{p_0}{p} \right)^{k_{21}} \right] = \lim_{r^{21} \to 0} \left[T \left(\frac{p_0}{p} \right)^{k_{21}} \right] = \left[T \left(\frac{p_0}{p} \right)^{k_{21}} \right]_{r^{21}=0} \qquad (12.28)$$

meaning that all the water vapor in the system either transforms to liquid water or ice before $T = 0$ has been reached. This we may deduce from (10.66)

$$\lim_{T \to 0} r^{21}(p, T) = \lim_{T \to 0} \frac{R_0 p^{21}(T)}{R_1 [p - p^{21}(T)]} = 0 \qquad (12.29)$$

Due to the lack of an exact analytical formulation of $p^{21}(T)$ for the entire temperature range, we verify (12.29) by using a numerical table. From the set of numbers as given in Table 12.1 we recognize that $p^{21}(T)$ goes to zero much faster than T itself. This is the reason that the fourth term on the left-hand side of (12.27) vanishes as $T \to 0$. From (12.27) we then obtain a first expression for C

$$\left[T \left(\frac{p_0}{p} \right)^{k_{21}} \right]_{r^{21}=0} \exp \left[k_{21} \ln \left(\frac{R_0}{R_1} \right) \right] = C \qquad (12.30)$$

Since $k_{21} \approx k_0$ we can think of (12.30) to be a dry adiabat which approaches the moist adiabat as $T \to 0$. In fact, $T(p_0/p)^{k_{21}}$ is a specific type of a potential temperature whose adiabat exponent is not k_0 but k_{21}. This is a very artificial system consisting of dry air and liquid water (ice) in which evaporation is forbidden to keep $m^{H_2O} = m^2 = $ constant. The qualitative behavior of the various adiabats is shown in Figure 12.1.

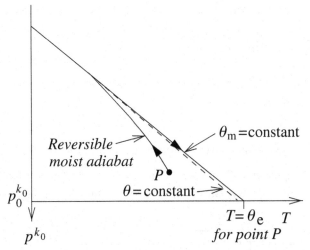

Fig. 12.1 Reversible moist adiabat, adiabats with exponents k_{21} and k_0, potential equivalent temperature.

Reference to (11.14) indicates that this particular system can be described by the adiabat equation

$$\left(c_{p,0}m^0 + c_2 m^2\right)\frac{dT}{T} - R_0 m^0 \frac{dp}{p} = 0 \tag{12.31}$$

It should be carefully noted that all water vapor has been condensed so that $m^{H_2O} = m^2$. Integrating this expression yields

$$T\left(\frac{p_0}{p}\right)^{k_{21}}_{r^{21}=0} = \text{constant}, \qquad k_{21} = \frac{R_0 m^0}{c_{p,0}m^0 + c_2 m^2} \tag{12.32}$$

The particular temperature which this adiabat assumes at $p = p_0$ is called the *potential equivalent temperature* θ_e so that from (12.30) we obtain

$$\theta_e = T\left(\frac{p_0}{p}\right)^{k_{21}}_{T\to 0} = C\exp\left[-k_{21}\ln\left(\frac{R_0}{R_1}\right)\right] \tag{12.33}$$

We now replace the constant C in (12.27) with the help of this expression and obtain the final version of the potential equivalent temperature.

$$\boxed{\theta_e = T\left(\frac{p_0}{p}\right)^{k_{21}}\exp\left[\frac{l_{21}k_{21}r^{21}}{R_0 T} + k_{21}\ln\left(1 + \frac{R_1}{R_0}r^{21}\right)\right]} \tag{12.34}$$

The term $\ln(1 + r^{21}R_1/R_0)$ in the exponent can be replaced with the help of (12.23). Inspection of (12.34) shows that θ_e is a constant of each particular system whereas in general θ_e is an intensive coordinate dependent on the variables (p, T, r^{21}). The graphical procedure of finding θ_e is shown in Figure 12.1. We follow the reversible moist adiabat until we approach absolute zero temperature.

Along this path all water vapor condenses so that the moist adiabat approaches the special adiabat with exponent k_{21}. Then we follow this adiabat, which is practically identical to the dry adiabat, down to $p = p_0$ where $T = \theta_e$. In Chapter 13, when we briefly discuss the thermodynamic diagrams, we shall return to graphical procedures. If the potential equivalent temperature of an ice cloud is needed we replace the indices 2 by 3.

12.4 Approximations to the reversible moist adiabat

The differential equation (12.26) describing the reversible moist adiabatic process is often stated in an approximate form. Approximating k_{21} in terms of k_0 we find

$$k_{21} = \frac{R_0}{c_{p,0}\left(1 + c_2 m^{H_2O}/c_{p,0}m^0\right)} \approx k_0\left(1 - \frac{c_2}{c_{p,0}}r^{H_2O}\right) \qquad (12.35)$$

With $d\ln T = d\ln \theta + k_0 d\ln p$ we obtain from (12.26) together with (12.35) the expression

$$d\ln\theta + \frac{k_0 c_2 r^{H_2O}}{c_{p,0}}d\ln p + d\left(\frac{l_{21}k_{21}r^{21}}{R_0T}\right) + k_{21}d\ln\left(\frac{R_0}{R_1} + r^{21}\right) = 0 \qquad (12.36)$$

In the literature this equation is also known as the *adiabatic differential equation*. The terms containing r^{H_2O} represent the condensation effect and $d\ln\theta$ solely describes the dry adiabatic change.

Formula (12.34), representing the potential equivalent temperature, can also be easily approximated by recognizing that $r^{21}R_1/R_0 \ll 1$ so that the term $\ln(1 + r^{21}R_1/R_0)$ may be neglected and by setting $k_{21} = k_0$ in the pressure term. This gives a first approximation to the potential equivalent temperature

$$\boxed{\theta_{e,1} \approx \theta \exp\left(\frac{l_{21}k_{21}r^{21}}{R_0T}\right)} \qquad (12.37a)$$

Since in this equation the exponent is a small number we easily find a second approximation to the potential equivalent temperature

$$\boxed{\theta_{e,2} \approx \theta\left(1 + \frac{l_{21}r^{21}}{c_{p,0}T}\right) = \left(\frac{p_0}{p}\right)^{k_0}\left(T + \frac{l_{21}r^{21}}{c_{p,0}}\right) = T_e^{21}\left(\frac{p_0}{p}\right)^{k_0}} \qquad (12.37b)$$

where k_{21} has been replaced by k_0. The quantity T_e^{21} is the *equivalent temperature of saturated air*. Some authors introduce an extra name for this approximation of θ_e.

Let us reconsider the first approximation of (12.37a) of θ_e. Introducing in this equation the additional simplification $k_{21} = k_0$ yields the formal definition of the *liquid water potential temperature* θ_1 as discussed in detail by Betts (1973)

$$\boxed{\theta_{e,3} = \theta_1 \approx \theta \exp\left(\frac{l_{21}k_0 r^{21}}{R_0 T}\right) = \theta \exp\left(\frac{l_{21} r^{21}}{c_{p,0} T}\right)} \qquad (12.37c)$$

This equation is the solution to the differential equation

$$\frac{d\theta_1}{\theta_1} = \frac{dT}{T} - k_0 \frac{dp}{p} + d\left(\frac{l_{21} r^{21}}{c_{p,0} T}\right) \qquad (12.38)$$

In order to determine to what extent θ_1 is a conservative quantity during a moist adiabatic process, we compare (12.38) with the differential equation (12.26) for the reversible moist adiabat. Replacing in (12.26) k_{21} by k_0 and ignoring the last term in this equation we find that $d\theta_1 = 0$. Thus we conclude that to a good approximation θ_1 remains constant in a moist adiabatic process.

12.5 The irreversible moist adiabat

In contrast to the moist reversible process which refers to a closed system, we will now consider the so-called *pseudoadiabatic process*. Such a process is characteristic of a system which is open for the mass of the condensed phase. Since a part of the thermal energy is lost as the condensation particles leave the system, we do not have a true adiabatic but only a pseudoadiabatic process. A special assumption is that the condensed phase will leave the system completely as soon as it forms. The moist reversible process then refers to the idealized atmosphere where clouds exist but no rain is falling out of these. In case of the special pseudoadiabatic process, rain falls to the ground but clouds do not exist. The truth will be somewhere in between these two idealized cases.

The assumptions leading to the derivation of the equation of the irreversible moist adiabat for a water cloud are summarized next. In case of an ice cloud the superscript $k = 2$ can be replaced by $k = 3$. In order to carry out the required mathematical manipulations, it is of advantage for physical insight to use extensive variables to begin with. Since the system is open for the condensed phase only, we have

$$\begin{aligned} d_e M^0 &= 0, & d_e M^1 &= 0, & d_e M^2 &\neq 0 \\ d_i M^1 &= dM^1, & d_i M^1 + d_i M^2 &= 0, & a_{21} &= 0 \end{aligned} \qquad (12.39)$$

We begin the analysis by repeating the extensive form of (11.21)

$$\left(c_{p,0} M^0 + c_{p,1} M^1 + c_2 M^2\right) \frac{dT}{T} - \left(R_0 M^0 + R_1 M^1\right) \frac{dp}{p} = l_{21} \frac{d_i M^2}{T} \qquad (12.40)$$

With the help of (10.39b), setting $v_2 = 0$, we find

$$\frac{l_{21}}{T} dM^1 = d\left(\frac{l_{21} M^1}{T}\right) + \frac{l_{21} M^1}{T^2} dT + \frac{(c_2 - c_{p,1})M^1}{T} dT \qquad (12.41)$$

Substituting (12.41) into (12.40) we obtain

$$\left(c_{p,0}M^0 + c_2 M^{H_2O}\right)\frac{dT}{T} - \left(R_0 M^0 + R_1 M^1\right)\frac{dp}{p} + d\left(\frac{l_{21}M^1}{T}\right) + \frac{l_{21}M^1}{T^2}\,dT = 0$$

(12.42)

In contrast to the moist reversible process (12.18), the quantity $M^{H_2O} = M^1 + M^2$ is no longer a constant of the system. We divide (12.42) by $c_{p,0}M^0$, which continues to be constant, and introduce the filter condition $a_{21} = 0$. We first observe that

$$\frac{M^1}{M^0} = \frac{m^1}{m^0} = r = r^{21}(p, T), \qquad \frac{M^{H_2O}}{M^0} = r^{21} + \frac{m^2}{m^0} = r^{H_2O}$$

(12.43)

This yields

$$\left(1 + \frac{c_2}{c_{p,0}}r^{H_2O}\right)\frac{dT}{T} - \frac{R_0}{c_{p,0}}\frac{dp}{p} - \frac{R_1 r^{21}}{c_{p,0}}\frac{dp}{p} + d\left(\frac{l_{21}r^{21}}{c_{p,0}T}\right) + \frac{l_{21}r^{21}}{c_{p,0}T^2}\,dT = 0 \quad (12.44)$$

Once again, it should be carefully noted that in this expression $r^{H_2O} = r^{H_2O}(p, T)$ is not a constant but a variable of the system. The last term on the left-hand side of (12.44) can be replaced by (12.25a). With this replacement we obtain

$$\boxed{\left(1 + \frac{c_2}{c_{p,0}}r^{H_2O}\right)d\ln T - k_0 d\ln p + d\left(\frac{l_{21}r^{21}}{c_{p,0}T}\right) + k_0 d\ln\left(\frac{R_0}{R_1} + r^{21}\right) = 0}$$

(12.45)

This is the equation of the irreversible moist adiabat which can be rewritten by introducing the potential temperature. The result is

$$d\ln\theta + \frac{c_2}{c_{p,0}}r^{H_2O}d\ln T + d\left(\frac{l_{21}r^{21}}{c_{p,0}T}\right) = -k_0 d\ln\left(\frac{R_0}{R_1} + r^{21}\right)$$

(12.46)

The quantity r^{21} is determined from the previously stated condition that the condensed phase leaves the system as soon as it forms so that

$$M^2 = 0, \qquad r^{H_2O} = r^{21}(p, T)$$

(12.47)

Equation (12.46) is the equation of the *pseudoadiabat* which can be integrated immediately yielding

$$\ln\theta + k_0\ln\left(\frac{R_0}{R_1} + r^{21}\right) + \frac{l_{21}r^{21}}{c_{p,0}T} + \frac{c_2}{c_{p,0}}\int_{T_0}^{T} r^{21}(p, T')d\ln T' = C \quad (12.48)$$

The constant of the system will now be evaluated. The special potential temperature which the system assumes when all water vapor has condensed, i.e. $r^{21} = 0$, is called the *pseudopotential temperature* θ_{ps}. This condition is realized only in the idealized case that $T \to 0$. According to Table 12.1 the third term drops out. Therefore, C can be replaced in terms of θ_{ps} so that (12.48) can be rewritten as

$$\ln\theta_{ps} + k_0\ln\left(\frac{R_0}{R_1}\right) + \frac{c_2}{c_{p,0}}\int_{T_0}^{0} r^{21}(p, T')d\ln T' = C$$

(12.49)

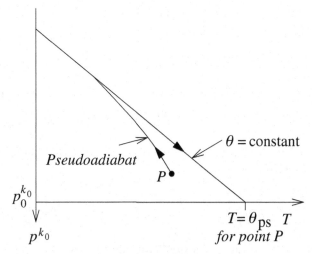

Fig. 12.2 Pseudoadiabat, pseudopotential temperature.

Replacing C in (12.48) by (12.49) we find

$$\ln\left(\frac{\theta_{ps}}{\theta}\right) = \frac{l_{21}r^{21}}{c_{p,0}T} + k_0 \ln\left(1 + \frac{R_1}{R_0}r^{21}\right) + \frac{c_2}{c_{p,0}}\int_0^T r^{21}(p, T')\,d\ln T' \quad (12.50)$$

With (12.23) we finally obtain

(a) $\quad \theta_{ps} = \theta \exp\left[\frac{l_{21}r^{21}}{c_{p,0}T} + k_0 \ln\left(1 + \frac{R_1}{R_0}r^{21}\right) + \frac{c_2}{c_{p,0}}\int_0^T r^{21}(p, T')\,d\ln T'\right]$

(b) $\quad \theta_{ps} = \theta \exp\left[\frac{l_{21}r^{21}}{c_{p,0}T} - k_0 \ln\left(\frac{p^0}{p}\right) + \frac{c_2}{c_{p,0}}\int_0^T r^{21}(p, T')\,d\ln T'\right]$

(c) $\quad \theta_{ps} = \theta \exp\left[\frac{l_{21}r^{21}}{c_{p,0}T} - k_0 \ln\left(1 - \frac{p^{21}}{p}\right) + \frac{c_2}{c_{p,0}}\int_0^T r^{21}(p, T')\,d\ln T'\right]$

$$(12.51)$$

In case of ice the superscript $k = 2$ must be replaced by $k = 3$. For a particular system, as follows from (12.49), $\theta_{ps} = $ constant since for a fixed T_0 the integral is simply a number. In general θ_{ps} is an intensive variable of state of the coordinates temperature T and pressure p since the mixing ratio r^{21} is a function of p and T.

Therefore, only one pseudoadiabat runs through a given point $P(p, T)$. This results from the equilibrium assumption $a_{21} = 0$ owing to the assumption that the condensed phase leaves the system as soon as it has formed. We find the value of the pseudoadiabat by the graphical procedure as shown schematically in Figure 12.2.

12.6 Construction of the pseudoadiabat on a thermodynamic chart

Many thermodynamic problems are solved with the help of thermodynamic charts or diagrams, some of these will be discussed in the next chapter. In order to treat the pseudoadiabatic process, lines of constant θ_{ps} are constructed with the help of (12.51). Due to the integral in the exponent we cannot evaluate this transcendental equation explicitly, but we must use an iterative technique which will now be outlined briefly.

Taking the logarithm of (12.51b), substituting for $\ln \theta$ and using (10.66) for the saturation mixing ratio, we obtain

$$\ln p^0 = \frac{\ln T}{k_0} + \frac{l_{21} p^{21}}{R_1 T p^0} + \frac{c_2}{R_1} \int_0^T \frac{p^{21}(T')}{p^0(T')} \, d \ln T' - \ln \left(\frac{\theta_{ps}^{1/k_0}}{p_0} \right) \qquad (12.52)$$

Next we introduce the following symbols for brevity:

$$C = \ln \left(\frac{\theta_{ps}^{1/k_0}}{p_0} \right), \qquad a(T) = \frac{l_{21} p^{21}(T)}{R_1 T}, \qquad b(T) = \frac{c_2 p^{21}(T)}{R_1} \qquad (12.53)$$

Instead of (12.52) we obtain

$$\ln p^0 = \frac{\ln T}{k_0} + \frac{a(T)}{p^0} + \int_0^T \frac{b(T')}{p^0(T')} \, d \ln T' - C \qquad (12.54)$$

It will be seen that p^0 is a function of temperature only. As stated before this transcendental equation is solved iteratively by using on the right-hand side previous values of $p^0(T)$. If ν counts the iteration step, then

$$\ln p^{0 \ (\nu+1)} = \frac{\ln T}{k_0} + \frac{a(T)}{p^{0 \ (\nu)}} + \int_0^T \frac{b(T')}{p^{0 \ (\nu)(T')}} \, d \ln T' - C \qquad (12.55)$$

The integral must be solved numerically. As a starting value for $\ln p^0$ we may set the integral equal to zero so that the right-hand side of (12.55) is known. Once $p^0(T)$ has been found iteratively, each pseudoadiabat $\theta_{ps} = $ constant can be constructed from the pressure–temperature relationship

$$p(T) = p^0(T) + p^{21}(T) \qquad (12.56)$$

and then plotted in the thermodynamic chart.

12.7 Dinkelacker's approximation of the moist adiabats

Equations (12.18) and (12.42) are the basic differential equations which lead to the formulation of the reversible and irreversible adiabats, respectively. Noting that

$r^{H_2O} \ll 1$, we may approximately set $r^{H_2O} c_2/c_{p,0} = 0$ in both equations which will then be identical. Introducing the filter condition $r = r^{21}(p, T)$, we obtain from (12.20) with $k_{21} \approx k_0$

$$d \ln T - k_0 d \ln p - \frac{R_1}{c_{p,0}} r^{21} d \ln p + d \left(\frac{l_{21} r^{21}}{c_{p,0} T} \right) + \frac{l_{21} r^{21}}{c_{p,0} T^2} dT = 0 \qquad (12.57)$$

The integration of this equation, using (12.25a) and the definition of the potential temperature, gives

$$\ln \theta = -\frac{l_{21} r^{21}}{c_{p,0} T} - k_0 \ln \left(\frac{R_0}{R_1} + r^{21} \right) + \ln C \qquad (12.58)$$

The constant C of the system is evaluated from the condition $r^{21} \to 0$ as $T \to 0$. The resulting special potential temperature is called θ_h so that

$$\ln C = \ln \theta_h + k_0 \ln \left(\frac{R_0}{R_1} \right) \qquad (12.59)$$

Substitution of this expression into (12.58) results in

$$\boxed{\theta_h = \theta \exp \left[\frac{l_{21} r^{21}}{c_{p,0} T} + k_0 \ln \left(1 + \frac{R_1}{R_0} r^{21} \right) \right]} \qquad (12.60)$$

Dinkelacker (1939) calls the constant of the system the *main adiabat* (German: *Hauptadiabate*) which is slightly less general than (12.51). If the logarithmic term is ignored in (12.60), since $r^{21} R_1/R_0 \ll 1$, we obtain an approximation to the main adiabat.

We conclude this chapter by making reference to the so-called *entropy potential temperature* which was introduced by Hauf and Höller (1987). The formula for this quantity was constructed in such a way that after the introduction of acceptable simplifications it reduced to various other types of potential temperatures as discussed in this section. Hauf and Höller also presented a prognostic equation for the entropy potential temperature which is similar to the heat equation that we have presented earlier. For more details the reader is invited to consult the original paper.

12.8 Problems

12.1: From the definition of the main adiabat find the moist adiabatic temperature gradient. Hint: First show that

$$1 + r^{21} \frac{R_1}{R_0} = \frac{p}{p - p^{21}}$$

Caution: Apply the required approximations near the end of the derivation instead of at the beginning.

12.2: Verify the identity (12.23).

13

Thermodynamic diagrams

13.1 General remarks

Thermodynamic charts or diagrams are used to provide graphical solutions to some of the processes which were described in the previous chapters. The diagrams contain isobars, isotherms, dry adiabats, pseudoadiabats, lines of constant saturation mixing ratio and auxiliary lines. We will be very brief in our discussion and omit the description of various auxiliary lines that are needed, for example, in the construction of the pressure–height curve of a particular sounding. A full discussion of thermodynamic diagrams can be found in various reference books. Our reference goes to an excellent manual entitled *Use of the Skew T-Log p Diagram in Analysis and Forecasting* which was published in 1961 by the United States Air Force.

Energy changes due to thermodynamic processes are of great importance in many meteorological considerations. Therefore, it is of primary importance that the area enclosed by lines on a particular diagram representing a cyclic process are equal to or at least proportional to the work done during such a process. The ordinary work diagram with coordinates (p, v) is not very suitable for meteorological applications since v is not an observed quantity. Therefore, it is desirable to construct diagrams with coordinates pressure and temperature, two quantities which are regularly observed. There are three major criteria by which the usefulness of a diagram may be established:

(i) Is the work done in a cyclic atmospheric process proportional to the area enclosed by the lines representing the process?

(ii) How large is the angle between isotherms and adiabats? A large angle is desirable since soundings drawn on the diagram are analyzed on the basis of their slopes. The larger the angle, the easier it is to observe changes in the slopes.

(iii) How many important isolines are straight? The more straight and the less curved are the lines, the easier it is to use the diagram.

13.2 Energy equivalence

A well-known theorem of analysis concerning the transformation of coordinates from the Cartesian (x, y)-system to an arbitrary (χ, ξ)-system is given by

$$\iint dx\, dy = \iint J\left(\frac{x, y}{\chi, \xi}\right) d\chi\, d\xi \quad \text{with}$$

$$J\left(\frac{x, y}{\chi, \xi}\right) = \frac{\partial x}{\partial \chi}\frac{\partial y}{\partial \xi} - \frac{\partial x}{\partial \xi}\frac{\partial y}{\partial \chi} \tag{13.1}$$

The quantity J is known as the Jacobian of the transformation or the functional determinant. The enclosed areas in the two coordinate systems are equal if and only if $J = 1$. If $J = $ constant then an enclosed area in the (x, y)-system will be uniformly expanded or contracted in the transformed coordinate system. This is an acceptable transformation. If J happens to involve either ξ or χ then the expansion or contraction is not uniform throughout the coordinate system.

We will now determine the Jacobians of three prominent thermodynamic diagrams. Afterwards sections of the diagrams and an example will be presented. Later we will have opportunities to refer to these diagrams as we discuss various concepts connected to thermodynamics.

13.2.1 Emagram

In this diagram the ordinate is $\ln p$ and the abscissa is the temperature T, see Figure 13.1. The pressure is normalized to standard pressure p_0. The isotherms and isobars are orthogonal to each other. Since the original diagram is (p, v), we set

$$x = v, \quad y = p \implies \chi = T, \quad \xi = \ln\left(\frac{p}{p_0}\right) \tag{13.2}$$

With the help of the ideal gas law for dry air we may write

$$x = \frac{R_0}{p_0}\chi \exp(-\xi), \quad y = p = p_0 \exp(\xi) \tag{13.3}$$

Taking the appropriate derivatives we find $J = R_0$.

13.2.2 Skew T-Log p diagram

In order to have a large angle between isotherms and adiabats, the isotherms will not be drawn perpendicularly to the isobars, but they will be inclined in the direction of

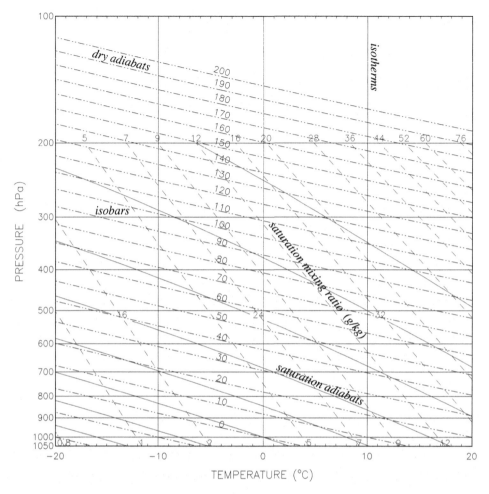

Fig. 13.1 The coordinate system of the emagram.

increasing temperature as shown in Figure 13.2. The transformation to be applied is

$$x = v, \quad y = p \implies \chi = T - K \ln\left(\frac{p}{p_0}\right), \quad \xi = \ln\left(\frac{p}{p_0}\right) \tag{13.4}$$

The constant $K > 0$ is chosen in such a way that the angle between isotherms and adiabats is nearly 90 degrees. With the help of the ideal gas law we find

$$x = \frac{R_0}{p_0}(\chi + k\xi)\exp(-\xi), \quad y = p = p_0\exp(\xi) \tag{13.5}$$

Calculation of the Jacobian gives $J = R_0$.

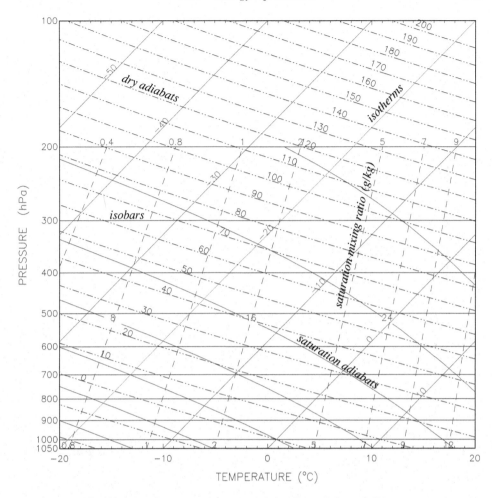

Fig. 13.2 The coordinate system of the Skew T-Log p diagram.

13.2.3 Stüve diagram

As in the emagram, isobars and isotherms are orthogonal, see Figure 13.3. We apply the following transformations:

$$x = v, \quad y = p \implies \chi = T, \quad \xi = \left(\frac{p}{p_0}\right)^{k_0} \tag{13.6}$$

This type of transformation makes it possible to include pressure $p = 0$ in the diagram which is not possible when pressure is drawn on a logarithmic scale. Application of the ideal gas law yields

$$x = \frac{R_0 \chi \xi^{-1/k_0}}{p_0}, \qquad y = p_0 \xi^{1/k_0} \tag{13.7}$$

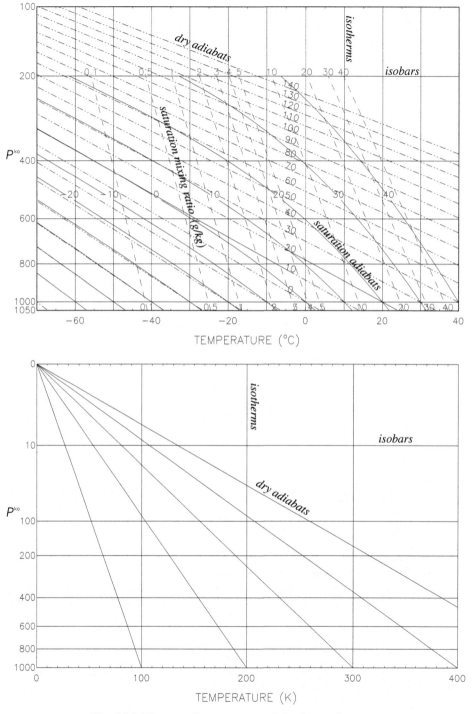

Fig. 13.3 The coordinate system of the Stüve diagram.

Table 13.1. *Correction factors for energy calculations with the Stüve diagram*

Pressure (hPa)	Correction factor $(p/p_0)^{k_0}$
1000	1.000
800	1.066
600	1.158
500	1.219
400	1.301
300	1.413

Computation of the Jacobian gives $J = c_{p,0}(p_0/p)^{k_0}$ which is a function of the ordinate p. The specific heat c_p of constant pressure has the same units as R_0. If $c_{p,0}$ is multiplied by an increment of temperature one obtains units of energy per unit mass. In order to carry out energy calculations with the Stüve diagram, correction factors must be applied which are listed in Table 13.1. This table covers the important meteorological pressure range. Despite the undesirable property of the Jacobian, the Stüve diagram is still used by the German Weather Service. The reason for this will be stated shortly.

13.3 Properties of the diagrams

A common feature of all diagrams is that the saturation adiabats are distinctly curved. We will now briefly discuss each diagram.

13.3.1 Emagram

A section of this diagram is shown in Figure 13.1. The dominant features of the emagram are listed next.

 (i) Area is proportional to energy.
 (ii) The angle between adiabats and isotherms is large enough to be reasonably convenient for computations.
 (iii) Isobars and isotherms are straight and perpendicular to each other. The adiabats are gently curved in the ordinary meteorological range. Saturation mixing ratio lines are straight.

13.3.2 Skew T-Log p diagram

A section of the diagram is shown in Figure 13.2. The essential features are listed next.

(i) Area is proportional to energy.
(ii) The angle between isotherms and adiabats is close to 90 degrees everywhere within the ordinary meteorological range. This feature makes the diagram very convenient for routine work.
(iii) Isobars and isotherms are straight but not orthogonal. Adiabats are gently curved, saturation mixing ratio lines are straight.

13.3.3 Stüve diagram

A section of the diagram is shown in Figure 13.3 as well as the complete frame. The main characteristics of the Stüve diagram are:

(i) Area is not proportional to energy. However, correction factors are provided so that the diagram can still be used for energy computations, see Table 13.1.
(ii) Isotherms and adiabats intersect at about the same angle as in the emagram which is roughly 45 degrees in the meteorological relevant range.
(iii) Isobars, isotherms and saturation mixing ratio are straight lines. A particular feature is that the dry adiabats are straight lines due to the choice of the ordinate. This makes the Stüve diagram a convenient tool for stability considerations when the slopes of a sounding and the dry adiabat are compared.

In Chapter 14 various applications of the diagrams will be presented. In the following section an example will be shown of how to use the Skew T-Log p diagram. The same rules apply to the other diagrams.

13.4 Graphical computation of the potential equivalent temperature

In Section 12.2 we have derived a formula for the equivalent temperature due to isobaric cooling that was also adiabatic. It is possible to obtain an approximate value of the equivalent temperature by a different process which is not isobaric. The reason that the isobaric equivalent temperature and the so-called adiabatic equivalent temperature T_e obtained from the diagram are numerically close is that the amount of latent heat of condensation which is released is nearly the same in both processes.

Consider a parcel of air characterized by the temperature and the dew point at the 700 hPa level as shown in Figure 13.4. An unsaturated parcel of air at the 700 hPa level is displaced dry adiabatically from point 1 in the upward direction. As the air cools due to the displacement, the water vapor within the parcel will be brought to condensation at a pressure level at which the saturation mixing ratio line going through the dew point temperature of the parcel (at point 1′) intersects the dry adiabat (point 2). From this point of intersection follow the saturation adiabat

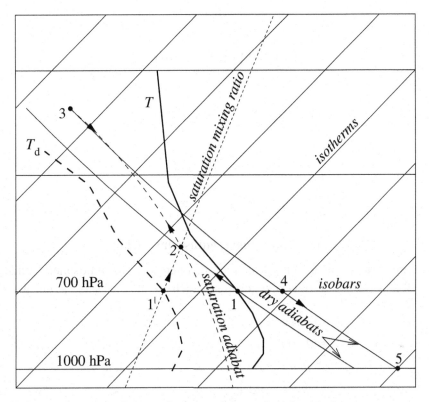

Fig. 13.4 Determination of the equivalent and the potential equivalent temperature using the Skew T-Log p diagram.

in the upward direction to a pressure level at which the saturation and the dry adiabats become parallel (point 3). At this level all the water vapor in the parcel has condensed out of the parcel. From point 3 follow the dry adiabat back to the original pressure level (point 4). The isotherm at the point of intersection at the 700 hPa level is equal to the equivalent temperature. The dry adiabat is now followed down to the 1000 hPa level (point 5). The isotherm through the point of intersection is the potential equivalent temperature. This completes the problem.

13.5 Problems

13.1: Verify the Jacobians of the emagram, the Skew T-Log p diagram and the Stüve diagram.

13.2: Assume that $T = 278$ K and $T_d = 269$ K at $p = 800$ hPa. Use a thermodynamic diagram of your choice to find the equivalent temperature of moist air. Compare your result using a suitable formula from the previous chapter.

14

Atmospheric statics

14.1 Atmospheric equiscalar surfaces

There are various scalar fields which are of interest to the atmospheric scientist. We will now list some equiscalar surfaces which are indispensable in our work.

isobaric surface :	p = constant
isothermal surface :	T = constant
isopicnic surface :	ρ = constant
isosteric surface :	v = constant
isentropic surface :	s = constant
geopotential surface :	ϕ = constant

These equiscalar surfaces are usually related in some form. For dry air the surfaces p, T and ρ are related by the ideal gas law. The equiscalar surfaces, in general, are functions of time. For example, the 500 hPa isobaric surface at a certain station today may have a height of 5000 m, some time later the height may be only 4800 m. In our present discussion we shall assume that all functions representing equiscalar surfaces refer to fixed time t_0. Furthermore, to keep things simple, the analysis will be restricted to the Cartesian frame. Thus we may write

$$\Psi = \Psi(\mathbf{r}, t_0) = \Psi(x^i, t_0), \qquad i = 1, 2, 3 \tag{14.1}$$

where $\mathbf{r} = x^1 \mathbf{i}_1 + x^2 \mathbf{i}_2 + x^3 \mathbf{i}_3$ is the position vector. Equiscalar surfaces of a certain type such as T = constant obviously cannot intersect for reasons of uniqueness, but in general they will not be parallel to one another. On the other hand, equiscalar surfaces representing various properties such as pressure and temperature will normally intersect. This results in the formation of the so-called *solenoidal tubes* as shown in Figure 14.1a.

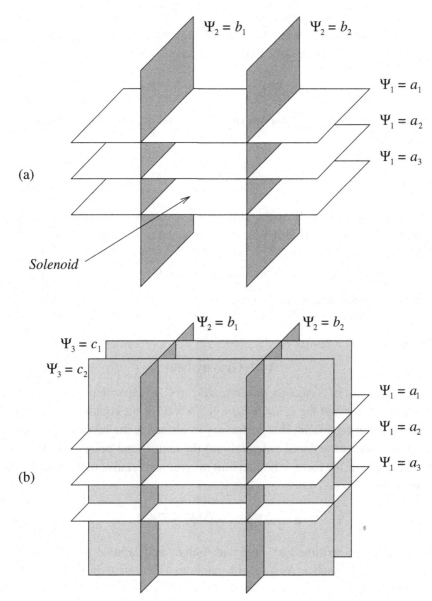

Fig. 14.1 Formation of solenoidal tubes (a) and cells (b) by intersecting scalar surfaces Ψ_1, Ψ_2 and Ψ_3.

In the special but highly interesting case that a mathematical relationship exists between the arbitrary field functions Ψ_1 and Ψ_2 in the form $\Psi_1 = \Psi_1(\Psi_2)$, the surfaces $\Psi_1 =$ constant and $\Psi_2 =$ constant are parallel and, therefore, solenoidal tubes cannot form. This particular situation, known as *homotropy*, will be discussed later in some detail.

In case of three intersecting surfaces *solenoidal cells* will form as shown in Figure 14.1b. Suppose there exists a relationship involving three field functions as exemplified by the ideal gas law. In this case we may write

$$F(\Psi_1, \Psi_2, \Psi_3) = 0 \quad \text{or} \quad \Psi_1 = \Psi_1(\Psi_2, \Psi_3) \tag{14.2}$$

where we have solved $F = 0$ for one of the variables, say Ψ_1. Forming the gradient of this expression we find

$$\nabla \Psi_1 = \frac{\partial \Psi_1}{\partial \Psi_2} \nabla \Psi_2 + \frac{\partial \Psi_1}{\partial \Psi_3} \nabla \Psi_3 \tag{14.3}$$

The three gradient vectors lie in the same plane so that the volume of the solenoidal cell is zero. This can be verified by forming the scalar triple product which vanishes, i.e.

$$[\nabla \Psi_1, \nabla \Psi_2, \nabla \Psi_3] = 0 \tag{14.4}$$

Therefore, a vanishing cell volume can also be interpreted as a relationship between the three field functions.

14.2 Gravity field

Gravity is the sum of *Newton's attractional force* $-\nabla \phi_a$ which is also called the *gravitational force* and the *centrifugal force* $-\nabla \phi_c$ of the rotating earth. In order to describe the gravity field, it is customary to introduce the *geopotential* ϕ which is the sum of the attractional potential and the centrifugal potential. The *gravity vector* **g** represents the force acting on unit mass at rest relative to the rotating earth and may be written as

$$\mathbf{g} = -\nabla \phi = -(\nabla \phi_a + \nabla \phi_c) \tag{14.5}$$

The potential ϕ is assumed to vary with height and latitude only. The smooth surface of the earth such as the sea on a calm day is assigned the potential zero

$$\phi_{\text{equator}} = \phi_{\text{pole}} = 0 \tag{14.6}$$

For most practical considerations the magnitude of gravity at the earth's surface may be approximated by

$$g = 9.780 + 0.021 \sin^2 \varphi + 0.031 \sin^2 \varphi \quad [\text{m s}^{-2}] \tag{14.7}$$

where φ refers to the geographical latitude. The first term refers to the value at the equator, the second term accounts for the deviation from the earth's spherical shape and the third term results from the centrifugal force. At the pole g exeeds the value at the equator by about 0.05 m s^{-2} so that very often we ignore the latitudinal variation altogether. If r_0 is the mean radius and M the mass of the earth, we may easily estimate the height variation of g from

$$g \approx \frac{\gamma M}{(r_0 + z)^2} = \frac{\gamma M}{r_0^2(1 + z/r_0)^2} \approx g(z = 0)\left(1 - \frac{2z}{r_0}\right) \qquad (14.8)$$

where $\gamma = 6.673 \times 10^{-11}$ N m^2 kg^{-2} is the *gravitational constant*. It turns out that at a height of 20 km the value of g is about 0.6% less than at sea level. Therefore, in the weather analysis of the troposphere and the lower stratosphere we are justified to treat g as height independent.

We now wish to calculate the *geopotential* ϕ for the elevation z at a fixed latitude. This leads to the introduction of the so-called *geopotential height* h to be expressed in *geopotential meters* (gpm). From (14.5) we find for a fixed latitude with $d(r_0 + z) = dz$

$$\phi = \int_{r_0}^{r_0+z} g(\varphi, z) \, dz = 9.8h \quad \text{[gpm]} \implies d\phi = 9.8 \, dh \qquad (14.9)$$

In the region of interest g varies primarily with latitude and less with height. Let us first consider the integral showing that ϕ can be interpreted as the *potential energy* imparted to unit mass if it is lifted from mean sea level to some height z. Surfaces of $\phi = $ constant are known as *level surfaces* which are surfaces of constant potential energy per unit mass. From mechanics we know that particles tend to move towards regions of minimum potential energy. If a particle is constrained to remain on a level surface it will have no inclination to slide along such a surface. Obviously this property is not shared by surfaces of constant elevation. The geopotential height introduced by (14.9) is in reality a measure of specific potential energy (m^2 s^{-2}). The dimensionless factor 9.8 was introduced for the purpose that the geometric and the geopotential meters have just about the same numerical value. For example, at the geographical latitude of 30° where $g = 9.793$ m s^{-2} a geopotential height of 3000 gpm corresponds to 3002 geometric meters.

Finally we should note that the difference between two *geopotential surfaces* located at heights z_1 and z_2 is given by

$$\phi_2 - \phi_1 = \int_{z_1}^{z_2} g(\varphi, z) \, dz \qquad (14.10)$$

The distance $z_2 - z_1$ between the two geopotential surfaces is smaller at the pole than at the equator since $g(90°) > g(0°)$.

14.3 Integration in equiscalar fields

We consider the spatial distribution of equiscalar surfaces of two arbitrary field functions Ψ_1 and Ψ_2. Our first task is to form the integral

$$\oint_\Gamma \Psi_1 d_g \Psi_2 = \oint_\Gamma d\mathbf{r} \cdot (\Psi_1 \nabla \Psi_2)$$
$$= \int_F d\mathbf{f} \cdot \nabla \times (\Psi_1 \nabla \Psi_2) = \int_F d\mathbf{f} \cdot (\nabla \Psi_1 \times \nabla \Psi_2) \qquad (14.11)$$

where $d_g \psi = d\mathbf{r} \cdot \nabla \psi$ is the *geometric differential* of ψ, that is the time-independent part of $d\psi$. Details are shown in Figure 14.2. The transformation from the line integral to the surface integral was accomplished by means of Stokes' integration theorem. The question arises quite naturally for which conditions the integral vanishes. Since the surface of integration F is arbitrary, the integral (14.11) vanishes if the cross-product of the two gradients is zero,

$$\nabla \Psi_1 \times \nabla \Psi_2 = 0 \qquad (14.12)$$

The particular situation that the integral (14.11) vanishes is known as the *condition of homotropy*. Obviously (14.12) is satisfied if both fields are completely homogeneous implying that both gradients vanish. Such a special condition is without interest and will be excluded from further considerations. Equation (14.12) is also satisfied by the less stringent conditions that one of the two gradients vanishes or if the two gradients are parallel or anti-parallel. In general, the distribution of atmospheric field variables is inhomogeneous so that the cross-product of the two gradients does not vanish and the integral in (14.11) is different from zero. This means that solenoids exist.

It is possible to rewrite (14.11) in the form

$$\oint_\Gamma \Psi_1 d_g \Psi_2 = \iint_F d\Psi_1 d\Psi_2 \qquad (14.13)$$

Details are given in textbooks on dynamical meteorology, see, for example, *DA*. This result is equivalent to the number of solenoids within the curve Γ as shown in Figure 14.2 for the (x, y)-plane. For this reason the nonzero vector formed by the cross-product

$$\boxed{\mathbf{N} = \nabla \Psi_1 \times \nabla \Psi_2} \qquad (14.14)$$

is defined as the *solenoidal vector*. From (14.14), noting that the divergence of the curl is zero, we find

$$\nabla \cdot \mathbf{N} = \nabla \cdot (\nabla \Psi_1 \times \nabla \Psi_2) = \nabla \cdot [\nabla \times (\Psi_1 \nabla \Psi_2)] = 0 \qquad (14.15)$$

showing that the solenoidal vector is divergence free.

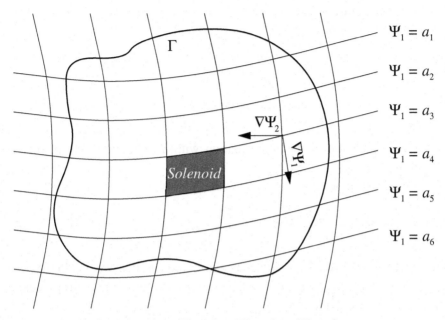

$$\Psi_1 = a_1$$
$$\Psi_1 = a_2$$
$$\Psi_1 = a_3$$
$$\Psi_1 = a_4$$
$$\Psi_1 = a_5$$
$$\Psi_1 = a_6$$

$$\Psi_2 = b_6 \quad \Psi_2 = b_5 \quad \Psi_2 = b_4 \quad \Psi_2 = b_3 \quad \Psi_2 = b_2 \quad \Psi_2 = b_1$$

Fig. 14.2 Integration in the field.

14.4 Homotropy, barotropy, heterotropy and piezotropy

14.4.1 Homotropy

Let us consider the case that the field function Ψ_1 is an indirect function of the position vector **r** so that

$$\Psi_1 = \Psi_1(\Psi_2(\mathbf{r}, t_0)) = \Psi_1(\Psi_2(x^i, t_0)) \qquad (14.16)$$

Only if Ψ_1 changes monotonically and continuously with Ψ_2 is it possible to invert this relation so that $\Psi_2 = \Psi_2(\Psi_1)$. Forming the geometric differential of (14.16) we obtain directly

$$
\begin{aligned}
d_g\Psi_1 &= \frac{\partial\Psi_1}{\partial\Psi_2}\frac{\partial\Psi_2}{\partial x^m}\,dx^m = \frac{\partial\Psi_1}{\partial\Psi_2}\left(\frac{\partial\Psi_2}{\partial x^1}\,dx^1 + \frac{\partial\Psi_2}{\partial x^2}\,dx^2 + \frac{\partial\Psi_2}{\partial x^3}\,dx^3\right) \\
&= \frac{\partial\Psi_1}{\partial\Psi_2}\,d\mathbf{r}\cdot\nabla\Psi_2 = \frac{\partial\Psi_1}{\partial\Psi_2}\,d_g\Psi_2 = d\mathbf{r}\cdot\nabla\Psi_1
\end{aligned}
\qquad (14.17a)
$$

and, therefore,

$$\frac{\partial\Psi_1}{\partial\Psi_2} = \frac{d\mathbf{r}\cdot\nabla\Psi_1}{d\mathbf{r}\cdot\nabla\Psi_2} = \frac{d_g\Psi_1}{d_g\Psi_2} \qquad (14.17b)$$

We may also find (14.17b) by carrying out the variation of $\Psi_1 = \Psi_1(\Psi_2)$ and replacing the variational δ by ∇. Thus we obtain

$$\delta\Psi_1 = \frac{\partial\Psi_1}{\partial\Psi_2}\delta\Psi_2, \qquad \delta = \nabla \implies$$

$$\nabla\Psi_1 = \frac{\partial\Psi_1}{\partial\Psi_2}\nabla\Psi_2 = \frac{d_g\Psi_1}{d_g\Psi_2}\nabla\Psi_2 \qquad (14.18)$$

Scalar multiplication of the last equation with $d\mathbf{r}$ completes the mathematical operation.

Vectorial multiplication of (14.18) with $\nabla\Psi_1$ gives

$$\nabla\Psi_1 \times \nabla\Psi_1 = \frac{d_g\Psi_1}{d_g\Psi_2}(\nabla\Psi_1 \times \nabla\Psi_2) = 0 \implies \nabla\Psi_1 \times \nabla\Psi_2 = 0 \qquad (14.19)$$

Thus surfaces of constant Ψ_1 and constant Ψ_2 are parallel. We would like to point out that relation (14.16) is just another way of stating the homotropy condition, which can also be written as

$$F(\Psi_1, \Psi_2) = 0 \qquad (14.20)$$

We recognize this by forming the geometric differential of (14.20). We find immediately

$$d_g F = d\mathbf{r} \cdot \nabla F = d\mathbf{r} \cdot \left(\frac{\partial F}{\partial\Psi_1}\nabla\Psi_1 + \frac{\partial F}{\partial\Psi_2}\nabla\Psi_2\right) = 0 \qquad (14.21)$$

and (14.12) follows. The relation (14.21) is used to define a quite useful quantity known as the *coefficient of homotropy* $\Gamma_{\Psi_2}^{\Psi_1}$. With reference to (14.17b) and (14.21) we obtain

$$\boxed{\Gamma_{\Psi_2}^{\Psi_1} = \frac{d_g\Psi_1}{d_g\Psi_2} = \frac{d\mathbf{r} \cdot \nabla\Psi_1}{d\mathbf{r} \cdot \nabla\Psi_2} = -\left(\frac{\partial F/\partial\Psi_2}{\partial F/\partial\Psi_1}\right)} \qquad (14.22)$$

The definition of this coefficient is fairly general since the functions Ψ_1 and Ψ_2 have not yet been specified.

14.4.2 Barotropy

A special case known as the *coefficient of barotropy* Γ_p^ρ is of particular interest and will be discussed next. Specifying in (14.12) Ψ_1 and Ψ_2 by the density ρ and the pressure p leads to the *condition of barotropy*

$$\boxed{\nabla\rho \times \nabla p = 0 \quad \text{or} \quad \rho = \rho(p) \quad \text{or} \quad F_g(\rho, p) = 0} \qquad (14.23)$$

From (14.22) we find for the coefficient of barotropy

$$\Gamma_p^\rho = \frac{d\mathbf{r} \cdot \nabla\rho}{d\mathbf{r} \cdot \nabla p} = \frac{d_g\rho}{d_g p} \tag{14.24}$$

Theoretical considerations frequently consider a special case of barotropy assuming that the density is constant throughout the medium. In this particular case the atmosphere is modeled by a rotating water tank. While the density gradient vanishes the pressure gradient differs from zero. For this special barotropic situation the coefficient of barotropy Γ_p^ρ is zero.

Finally it should be recognized that the state of the atmosphere over extended regions and periods of time is rarely barotropic. Nevertheless, the first successful numerical models for short range weather forecasts are based on this assumption. Next we will introduce some theory in connection with (14.23).

14.4.2.1 Consequences of the condition of barotropy

We wish to state a very interesting consequence of the barotropy condition $\rho = \rho(p)$. If $\nabla_h p$ is the horizontal pressure gradient any two of the three conditions

$$\text{(a)} \qquad \nabla\rho \times \nabla p = 0$$

$$\text{(b)} \qquad \frac{\partial}{\partial z}\left(\frac{1}{\rho}\nabla_h p\right) = 0 \tag{14.25}$$

$$\text{(c)} \qquad \frac{1}{\rho}\frac{\partial p}{\partial z} = f(z)$$

imply the correctness of the third one. To prove this we first obtain some auxiliary relations from conditions (14.25b) and (14.25c). Carrying out the operation (14.25b) we find

$$\text{(a)} \qquad \frac{\partial}{\partial z}\left(\frac{1}{\rho}\nabla_h p\right) = -\frac{1}{\rho^2}\frac{\partial\rho}{\partial z}\nabla_h p + \frac{1}{\rho}\nabla_h\frac{\partial p}{\partial z} = 0 \quad \text{or}$$

$$\text{(b)} \qquad -\frac{1}{\rho^2}\frac{\partial\rho}{\partial z}\nabla_h p + \nabla_h\left(\frac{1}{\rho}\frac{\partial p}{\partial z}\right) + \frac{1}{\rho^2}\frac{\partial p}{\partial z}\nabla_h\rho = 0 \implies \tag{14.26}$$

$$\text{(c)} \qquad -\frac{\partial\rho}{\partial z}\nabla_h p + \frac{\partial p}{\partial z}\nabla_h\rho = 0$$

The second term of (14.26b) vanishes according to (14.25c) since the differentiation implied by the horizontal gradient is with respect to x and y. Vectorial multiplication of (14.26c) with the horizontal density gradient gives

$$\nabla_h p \times \nabla_h \rho = 0 \tag{14.27}$$

which is analogous to condition (14.25a).

Taking the curl of the three-dimensional pressure force, using condition (14.25a), yields

$$\nabla \times \left(\frac{1}{\rho} \nabla p \right) = -\frac{1}{\rho^2} \nabla \rho \times \nabla p = 0 \tag{14.28}$$

Since the curl of the gradient of any function vanishes, we may replace the pressure gradient force by

$$\frac{1}{\rho} \nabla p = \nabla P \tag{14.29a}$$

Scalar multiplication of (14.29a) by the vertical unit vector gives

$$\frac{1}{\rho} \frac{\partial p}{\partial z} = \frac{\partial P}{\partial z} \tag{14.29b}$$

and consequently

$$\frac{1}{\rho} \nabla_h p = \nabla_h P \tag{14.29c}$$

Now we have derived the required auxiliary relations.

First we wish to show that (14.25b) follows from the validity of (14.25a) and (14.25c). We note that (14.29) resulted from condition (14.25a). Carrying out the vertical differentiation of (14.29c) and then using condition (14.25c) gives

$$\frac{\partial}{\partial z} \left(\frac{1}{\rho} \nabla_h p \right) = \frac{\partial}{\partial z} (\nabla_h P) = \nabla_h \left(\frac{\partial P}{\partial z} \right) = \nabla_h \left(\frac{1}{\rho} \frac{\partial p}{\partial z} \right) = \nabla_h f(z) = 0 \tag{14.30}$$

Now we show that application of (14.25a) and (14.25b) implies (14.25c). Instead of using (14.25a) directly we use (14.29c) which followed from (14.25a). Vertical differentiation of (14.29c), upon using (14.25b), gives condition (14.25c).

$$\frac{\partial}{\partial z} \left(\frac{1}{\rho} \nabla_h p \right) = \nabla_h \left(\frac{\partial P}{\partial z} \right) = \nabla_h \left(\frac{1}{\rho} \frac{\partial p}{\partial z} \right) = 0 \implies \frac{1}{\rho} \frac{\partial p}{\partial z} = f(z) \tag{14.31}$$

The last step becomes obvious by recalling that the differentiation implied by the horizontal gradient operator is with respect to x and y.

Finally, we must show that conditions (14.25b) and (14.25c) imply (14.25a). Splitting the gradient operators appearing in (14.25a) into their horizontal and vertical parts, using the auxiliary conditions (14.26c) and (14.27) which resulted from (14.25b,c) gives the required result

$$\nabla \rho \times \nabla p = \left(\nabla_h \rho + \mathbf{k} \frac{\partial \rho}{\partial z} \right) \times \left(\nabla_h p + \mathbf{k} \frac{\partial p}{\partial z} \right)$$

$$= \nabla_h \rho \times \nabla_h p + \mathbf{k} \times \left(\frac{\partial \rho}{\partial z} \nabla_h p - \frac{\partial p}{\partial z} \nabla_h \rho \right) = 0 \tag{14.32}$$

14.4.2.2 Local conservation of barotropy

The question now arises if there exist conditions which guarantee the local conservation of barotropy so that an originally barotropic atmosphere remains in this particular state. Taking the local time derivative of the condition of barotropy we obtain

$$
\frac{\partial}{\partial t}(\nabla \rho \times \nabla p) = \nabla \left(\frac{\partial \rho}{\partial t}\right) \times \nabla p + \nabla \rho \times \nabla \left(\frac{\partial p}{\partial t}\right)
$$

$$
= \nabla \left(\frac{d\rho}{dt} - \mathbf{v} \cdot \nabla \rho\right) \times \nabla p + \nabla \rho \times \nabla \left(\frac{dp}{dt} - \mathbf{v} \cdot \nabla p\right)
$$

(14.33)

where we involved the total derivative with the help of the Eulerian expansion. The initial barotropy at time $t = t_0$ is characterized by the coefficient of barotropy. Utilizing (14.24) and observing that in general $d\mathbf{r} \neq 0$, we may write

$$
d\mathbf{r} \cdot \left(\Gamma_p^\rho \nabla p - \nabla \rho\right) = 0 \implies \nabla \rho = \Gamma_p^\rho \nabla p
$$

(14.34)

Substituting (14.34) into (14.33) we find after some obvious steps

$$
\frac{\partial}{\partial t}(\nabla \rho \times \nabla p) = \nabla \left(\frac{d\rho}{dt} - \Gamma_p^\rho \mathbf{v} \cdot \nabla p\right) \times \nabla p - \nabla \left(\Gamma_p^\rho \frac{dp}{dt} - \Gamma_p^\rho \mathbf{v} \cdot \nabla p\right) \times \nabla p
$$

$$
+ \left(\frac{dp}{dt} - \mathbf{v} \cdot \nabla p\right) \nabla \Gamma_p^\rho \times \nabla p
$$

(14.35)

The third term on the right-hand side must vanish for the following reason. In barotropy, in general, the density is a function of pressure, so that the coefficient of barotropy is also a function of the pressure p. Therefore, the gradient of the coefficient of barotropy is proportional to the pressure gradient so that the vector product vanishes. The remaining part of (14.35) reduces to

$$
\frac{\partial}{\partial t}(\nabla \rho \times \nabla p) = \nabla \left(\frac{d\rho}{dt} - \Gamma_p^\rho \frac{dp}{dt}\right) \times \nabla p
$$

(14.36)

The expression on the right-hand side vanishes if

$$
\frac{d\rho}{dt} = \Gamma_p^\rho \frac{dp}{dt}
$$

(14.37)

Whenever the density and pressure fields are related in this way the barotropy existing locally at time $t = t_0$ will be maintained.

For the purpose of further discussion of the concept of local conservation of barotropy, (14.37) will be developed assuming that the atmosphere can be modeled by a frictionless gas consisting of dry air only. From the first law of thermodynamics as given by (3.36) we find

$$
\text{(a)} \quad \frac{dq}{dt} = c_{v,0}\frac{dT}{dt} + p\frac{dv}{dt}
$$
$$
\text{(b)} \quad \frac{dq}{dt} = c_{p,0}\frac{dT}{dt} - v\frac{dp}{dt}
$$

(14.38)

Multiplying (14.38a) by $c_{p,0}$ and (14.38b) by $c_{v,0}$ and then subtracting the resulting equations gives

$$
\frac{d\rho}{dt} = \frac{\rho}{\kappa p}\frac{dp}{dt} - \frac{\kappa - 1}{\kappa}\frac{\rho^2}{p}\frac{dq}{dt} \quad \text{with}
$$
$$
\kappa = \frac{c_{p,0}}{c_{v,0}}, \qquad \frac{\kappa - 1}{\kappa} = \frac{R_0}{c_{p,0}} = k_0
$$

(14.39)

Substitution of (14.39) into (14.36) yields various terms involving gradient expressions. Due to barotropy three of these expressions are functions of pressure only

$$
\nabla\left(\frac{\rho}{\kappa p}\right) = \nabla f_1(p) = \frac{df_1}{dp}\nabla p
$$
$$
\nabla\left(\frac{\rho^2}{p}\right) = \nabla f_2(p) = \frac{df_2}{dp}\nabla p
$$
$$
\nabla\Gamma_p^\rho = \nabla f_3(p) = \frac{df_3}{dp}\nabla p
$$

(14.40)

Since (14.36) involves a vectorial multiplication with the pressure gradient we obtain

$$
\frac{\partial}{\partial t}(\nabla\rho \times \nabla p) = \left[\left(\frac{\rho}{\kappa p} - \Gamma_p^\rho\right)\nabla\left(\frac{dp}{dt}\right) - \frac{\kappa - 1}{\kappa}\frac{\rho^2}{p}\nabla\left(\frac{dq}{dt}\right)\right] \times \nabla p \quad (14.41)
$$

The barotropy existing initially will be maintained if the right-hand side of this equation vanishes. This is the case if one of the following two situations occurs:

(i) Any atmospheric changes occur adiabatically, $dq/dt = 0$, and additionally either one of two cases occurs:

   (a) $\Gamma_p^\rho = \rho/\kappa p$ — this situation is known as *autobarotropy*, or if
   (b) $\Gamma_p^\rho \neq \rho/\kappa p$. Then we must assume that $dp/dt = f(p)$ so that $\nabla(dp/dt) \times \nabla p = 0$.

(ii) If the atmospheric change does not take place adiabatically, $dq/dt \neq 0$, then in addition to (a) and (b) we must require that dq/dt is a function of pressure only. We will return to the condition of autobarotropy shortly.

14.4.3 Heterotropy

Whereas the condition of homotropy could be stated in the form (14.20), the requirement of heterotropy is more complicated since it involves three arbitrary field functions

$$F_g(\Psi_1, \Psi_2, \Psi_3) = 0 \quad \text{or} \quad \Psi_1 = \Psi_1(\Psi_2, \Psi_3) \tag{14.42}$$

where we have solved for one of these, say Ψ_1. Taking the gradient of this expression according to (14.3) we obtain

$$\nabla\Psi_1 = \left(\frac{\partial\Psi_1}{\partial\Psi_2}\right)_{\Psi_3} \nabla\Psi_2 + \left(\frac{\partial\Psi_1}{\partial\Psi_3}\right)_{\Psi_2} \nabla\Psi_3 \tag{14.43}$$

Successive multiplication of this expression with the gradients of the three field functions results in

$$
\begin{aligned}
&\text{(a)} \quad \nabla\Psi_1 \times \nabla\Psi_2 \neq 0 \\
&\text{(b)} \quad \nabla\Psi_1 \times \nabla\Psi_3 \neq 0 \\
&\text{(c)} \quad \nabla\Psi_2 \times \nabla\Psi_3 \neq 0
\end{aligned}
\tag{14.44}
$$

showing the existence of solenoidal tubes. Solenoidal cells, however, cannot exist. For example, by substituting (14.43) into (14.44b) we can see that relation (14.44b) is proportional to (14.44c).

A particularly interesting and important example is the *baroclinicity condition*. Let $\Psi_1 = \rho$ and $\Psi_2 = p$. According to (14.44a) the solenoidal vector differs from zero so that the equiscalar surfaces of pressure and density are inclined to each other. The baroclinicity condition may then be stated by

$$\nabla\rho \times \nabla p \neq 0, \qquad F_g(\rho, p, \Psi_3) = 0, \qquad \rho = \rho(p, \Psi_3) \tag{14.45}$$

In general, the atmosphere exhibits a baroclinic structure. A detailed discussion of *baroclinic instability* of atmospheric processes is given in textbooks on atmospheric dynamics, see, for example, *DA*.

14.4.4 Piezotropy

In contrast to the conditions of homotropy and heterotropy which refer to geometric changes, the concept of piezotropy has reference to changes experienced by individual air parcels. The relation between pressure and some field function χ is then given by

$$F(\chi, p) = 0 \tag{14.46}$$

Analogously to the barotropy coefficient (14.24), the *coefficient of piezotropy* is defined by

$$\gamma_p^\chi = \frac{d\chi}{dp} \tag{14.47}$$

An incompressible fluid, for example, is characterized by

$$\gamma_p^\rho = \frac{d\rho}{dp} = 0 \tag{14.48}$$

since for incompressibility the individual change $d\rho = 0$. The geometric change $d_g\rho$, however, may differ from zero for a multi-component system since the density gradient does not vanish. We have discussed this situation in Section 1.3.2 where we considered a salt solution.

An interesting example is the coefficient of piezotropy for dry air. A little reflection will show that for an adiabatic change (14.39) can be rewritten as

$$\frac{dp}{p} - \kappa\frac{d\rho}{\rho} = 0 \quad \text{or} \quad p\rho^{-\kappa} = \text{constant} \tag{14.49}$$

so that this piezotropy coefficient can be stated as

$$\boxed{\gamma_p^\rho = \frac{1}{\kappa R_0 T} = \frac{1}{c_L^2}} \tag{14.50}$$

where c_L is the *Laplace speed of sound*. This relation is helpful in numerical weather prediction in connection with the elimination of sound waves from atmospheric prognostic systems. From the numerical point of view such waves are considered as noise. For an incompressible fluid, according to (14.48), the coefficient of piezotropy is zero so that $c_L = \infty$.

14.4.5 Autobarotropy

A sufficient though not necessary condition for autobarotropy is the identity of geometric and individual changes of the density with pressure that an air parcel experiences during an arbitrary displacement. According to (14.24) and (14.47) we then have

$$\frac{d_g\rho}{d_g p} = \frac{d\rho}{dp} \quad \text{or} \quad \Gamma_p^\rho = \gamma_p^\rho \tag{14.51}$$

An easily understood consequence of this requirement is that barotropy will be maintained at all times since in case of autobarotropy individual processes cannot influence the barotropic field. This can be written as

$$\frac{\partial}{\partial t}(\nabla\rho \times \nabla p) = 0 \tag{14.52}$$

An example will make this clearer. Suppose that the atmosphere is dry adiabatically stratified. If an air parcel is also displaced dry adiabatically then we have autobarotropic conditions since the given density distribution as a function of pressure $\rho = \rho(p)$ cannot be changed by such a process.

14.5 Hydrostatic equilibrium

We will now describe one of the most important concepts in all of meteorology, known as the hydrostatic equilibrium. This leads to the derivation of the hydrostatic equation which will be obtained from energy transformations. We consider an atmospheric system which is closed with respect to mass and energy. The energy content within the fluid volume $V(t)$ consists of three parts

$$E_{\text{kin}} + E_{\text{pot}} + E_{\text{int}} = \text{constant} \quad \text{with}$$

$$E_{\text{kin}} = \int_{V(t)} \rho \frac{\mathbf{v}^2}{2} \, d\tau, \qquad E_{\text{pot}} = \int_{V(t)} \rho \phi \, d\tau, \qquad E_{\text{int}} = \int_{V(t)} \rho e \, d\tau \tag{14.53}$$

As usual, ρ is the density, ϕ the geopotential and e the specific *internal energy*. The sum of $E_{\text{pot}} + E_{\text{int}}$ is sometimes denoted as the *total potential energy* of the fluid volume. Since the total energy in such a system is conserved, we find that an increase of the *kinetic energy* must be associated with a decrease of the total *potential energy*. This process is expressed by

$$\frac{dE_{\text{kin}}}{dt} = -\frac{d}{dt}(E_{\text{pot}} + E_{\text{int}}) \tag{14.54}$$

According to *Dirichlet's stability theorem* the fluid volume will be in hydrostatic equilibrium if the total potential energy is characterized by an extremum

$$\boxed{\delta(E_{\text{pot}} + E_{\text{int}}) = 0} \tag{14.55}$$

Stable equilibrium exists if the total potential energy of the fluid volume has a minimum value while instability will be observed if it has reached a maximum value. Substituting (14.53) into (14.55) and setting $\delta = d/dt$ we find

$$\frac{d}{dt}(E_{\text{pot}} + E_{\text{int}}) = \int_{V(t)} \rho \left(\frac{d\phi}{dt} + \frac{de}{dt} \right) d\tau = 0 \tag{14.56}$$

where the differentiation rules for fluid volume integrals have been applied.

Now we need to set up the proper expressions for the time changes of the geopotential and the specific internal energy. In the region of meteorological interest we may represent the geopotential with sufficient approximation by a linear function of height. Using the Euler expansion for the geopotential we find

$$\frac{d\phi}{dt} = \left(\frac{\partial\phi}{\partial t}\right)_z + \mathbf{v} \cdot \nabla\phi = w\frac{\partial\phi}{\partial z} = wg \qquad \text{with} \quad \phi = gz + \text{constant} \quad (14.57)$$

Obviously the time derivative of the geopotential at a fixed position must vanish. The time change of the specific internal energy will be calculated with the help of the first law of thermodynamics. Ignoring frictional and radiative effects we obtain from (3.33) with the help of the continuity equation (2.12)

$$\frac{de}{dt} = -v\nabla \cdot \mathbf{J}^h - v\nabla \cdot (p\mathbf{v}) + v\mathbf{v} \cdot \nabla p \qquad (14.58)$$

Substituting (14.57) and (14.58) into (14.56) and applying Gauss' integral theorem, we find

$$\int_{V(t)} \rho\mathbf{v} \cdot (\nabla\phi + v\nabla p)\, d\tau = \int_{F(t)} (\mathbf{J}^h + p\mathbf{v}) \cdot d\mathbf{f}$$

$$\mathbf{J}^h \cdot d\mathbf{f} = 0 \quad \text{thermodynamically closed} \qquad (14.59)$$

$$p\mathbf{v} \cdot d\mathbf{f} = 0 \quad \text{mechanically closed}$$

The right-hand side of this equation is zero since we have assumed a completely closed system thus ruling out any energy and mass transfer and interaction with the surroundings. Since no restrictions on the volume and the velocity were assumed, we have as the condition of hydrostatic equilibrium

$$\boxed{\nabla\phi + v\nabla p = 0} \qquad (14.60)$$

This equation describes the equilibrium state that the system assumes in the absence of kinetic energy transformations, that is, no rearrangement of air masses takes place. Decomposition of the three-dimensional gradient vectors into their horizontal and vertical components, and in view of (14.57), yields the following two requirements for hydrostatic equilibrium

$$
\boxed{
\begin{array}{ll}
\text{(a)} & \nabla_h\phi = -v\nabla_h p = 0 \\[2mm]
\text{(b)} & \dfrac{dp}{dz} = -g\rho \implies dp = -\rho\,d\phi = -9.8\rho\,dh
\end{array}
} \qquad (14.61)
$$

Equation (14.61b) is known as the *hydrostatic equation* where h, as usual, denotes the geopotential height. One word of caution. In meteorological practice and theory one often speaks somewhat loosely of hydrostatic equilibrium if condition (14.61b) alone is fullfilled even though condition (14.60) is not satisfied. In this case we really speak of the *hydrostatic approximation*.

We will now investigate the consequences of the hydrostatic equilibrium (14.60) by considering this statement in its various equivalent forms. In all cases the hydrostatic equilibrium is represented by two gradient expressions. By taking the curl of each expression, as shown in (14.62), only one term remains which can be easily interpreted. Using the equation of state for dry air we obtain

(a) $\qquad \nabla\phi + v\nabla p = 0,$ $\qquad\qquad \nabla\rho \times \nabla p = 0 \implies$ $\qquad\qquad \rho = \rho(p)$

(b) $\quad \nabla\phi + R_0 T\nabla \ln p = 0,$ $\qquad \dfrac{R_0}{p}\nabla T \times \nabla p = 0 \implies$ $\qquad\qquad p = p(T)$

(c) $\qquad \nabla\phi + \theta\nabla\Pi = 0,$ $\qquad\qquad \Pi = c_{p,0}\left(\dfrac{p}{p_0}\right)^{k_0},\qquad k_0 = \dfrac{R_0}{c_{p,0}}$

$\qquad\qquad \nabla\theta \times \nabla\Pi = 0 \implies$ $\qquad\qquad \theta = \theta(\Pi)\quad \text{or}\qquad \theta = \theta(p)$

(d) $\qquad \nabla\phi + v\nabla p = 0,$ $\qquad\qquad \nabla\rho \times \nabla\phi = 0 \implies$ $\qquad\qquad \rho = \rho(\phi)$

(e) $\qquad\qquad p = p(\phi),$ $\qquad\qquad \theta = \theta(\phi)$

$$\text{(14.62)}$$

Part (a) shows that the hydrostatic condition implies *barotropy*. Inverting this statement is not possible since barotropy does not require strict hydrostatic equilibrium. Part (b) shows that in hydrostatic equilibrium the pressure and the temperature fields are homotropically adjusted. In part (c) we have introduced the *Exner function* Π as shorthand notation. We find that in strict hydrostatic equilibrium the potential temperature depends on pressure only. Part (d) shows that in hydrostatic equilibrium the density can be expressed as a function of the geopotential only. This situation is known as *potentiotropy*. This means that surfaces of constant density are parallel to equipotential surfaces. Finally, combining barotropy and potentiotropy we find the conditions listed in part (e). Summarizing, in strict hydrostatic equilibrium all equiscalar surfaces are parallel and can be obtained by renumbering the geopotential surfaces, that is, we have perfect homotropic conditions. Two very meaningful examples involving the hydrostatic equation will now be given.

Example 1   Find the barotropic coefficient Γ_p^ρ in case of hydrostatic equilibrium.

First we combine (14.61b) and (14.24). By eliminating the density differential with the help of the ideal gas law for moist air we obtain after some obvious steps

$$\Gamma_p^\rho = \frac{d_g\rho}{d_g p} = -\frac{1}{9.8\rho}\frac{d_g\rho}{dh} = \frac{1}{9.8dh}\left[\frac{d_g T_v}{T_v} - \frac{d_g p}{p}\right]$$
$$= \frac{1}{9.8T_v}\left[\frac{9.8}{R_0} - \left(-\frac{d_g T_v}{dh}\right)\right]$$

$$\text{(14.63)}$$

Introducing the *lapse rate of the virtual temperature*

$$\gamma_{v,g} = -\frac{d_g T_v}{dh} \tag{14.64}$$

we find

$$\Gamma_p^\rho = \frac{1}{9.8 T_v}\left(\frac{9.8}{R_0} - \gamma_{v,g}\right) \tag{14.65}$$

The knowledge of the virtual temperature distribution is then sufficient to calculate the coefficient of barotropy. In a homogeneous atmosphere of constant density this coefficient is zero so that

$$\gamma_{v,g} = \gamma_h = \frac{9.8}{R_0} = 3.41\,\text{K}/100\,\text{gpm} \quad \text{with} \quad R_0 = 287.05\,\text{m}^2\,\text{s}^{-2}\,\text{K}^{-1} \tag{14.66}$$

This is the largest lapse rate possible and still maintaining hydrostatic equilibrium. γ_h is often called the *autoconvective lapse rate*. If the existing atmospheric lapse rate exceeds γ_h, the density increases with height so that overturning will occur.

Example 2   Derive the barometric height formula.

Substituting the ideal gas law for moist air into the hydrostatic equation (14.61b) gives

$$\frac{d_g \ln p}{dh} = -\frac{9.8}{R_0 T_v} = -\frac{1}{H_p} \tag{14.67}$$

The quantity $H_p(h)$ which is introduced here, is known as the *scale height of pressure*. $H_p(h)$ is height dependent since the virtual temperature depends on height. Moreover, this equation reveals that the pressure in warm air decreases more slowly with height than in cold air. We can also interpret H_p as the *height of the homogeneous atmosphere* in which the density does not vary at all. In this case $T_v(h)$ is assigned the ground value $T_v(0)$. Integration of (14.67) between pressure levels $p(h_1)$ and $p(h_2)$ gives

$$\boxed{p_2 = p_1 \exp\left[-\frac{9.8}{R_0}\int_{h_1}^{h_2}\frac{1}{T_v}\,dh\right] = p_1 \exp\left[-\frac{9.8}{R_0}\frac{h_2 - h_1}{T_{v,m}}\right]} \tag{14.68a}$$

where the *barometric mean temperature*

$$T_{v,m} = \frac{h_2 - h_1}{\int_{h_1}^{h_2}(1/T_v)\,dh} = \int_{p_1}^{p_2}\frac{T_v}{\ln(p_2/p_1)}\,d\ln p \tag{14.68b}$$

has been introduced. The last expression in this equation was obtained with the help of (14.67) and (14.68a). Equation (14.68a) is known as the *barometric height formula*. The barometric mean temperature represents a harmonic mean which is slightly less than the arithmetic mean temperature $T_{v,a}$ defined by

$$T_{v,a} = \frac{1}{h_2 - h_1} \int_{h_1}^{h_2} T_v dh \geq T_{v,m} \qquad (14.69)$$

If the averaging interval $(h_2 - h_1)$ is not too large, we may assume that the two mean temperatures are equal. For practical purposes we usually find the barometric mean temperature with the help of a thermodynamic chart.

14.6 Polytropic atmospheres

An atmosphere is called polytropic whenever the lapse rate of the virtual temperature is constant with height or

$$\gamma_{v,g} = -\frac{d_g T_v}{dh} = \text{constant} \qquad (14.70)$$

Integration of (14.70), proceeding from some reference level h_0 gives

$$T_v(h) = T_{v,0} - \gamma_{v,g}(h - h_0) \quad \text{with} \quad T_{v,0} = T_v(h_0) \qquad (14.71)$$

In order to accurately calculate the pressure–height curve we subdivide the atmosphere into a number of sufficiently thin sections and specify (14.71) for each of these. We may then substitute this equation into the barometric height formula (14.68a) and thus find the pressure change for any atmospheric subsection. By adding all pressure changes we obtain the required pressure–height curve. The same concept may also be used to construct polytropic model atmospheres. We shall refrain from going into details but we will show how to proceed.

The height H of the polytropic atmosphere is found from the requirement $T(H) = 0$. Selecting the surface of the earth as the reference level $h_0 = 0$, we find H from

$$H = \frac{T_{v,0}}{\gamma_{v,g}} \qquad (14.72)$$

Substitution of (14.72) into (14.71) gives the temperature distribution of the polytropic model atmosphere

$$T_v(h) = T_{v,0}\left(1 - \frac{h}{H}\right) \qquad (14.73)$$

The corresponding pressure–height distribution is found by substituting (14.71) into the barometric height formula (14.68a). A simple integration gives

$$p(h) = p_0 \left(\frac{T_{v,0} - \gamma_{v,g}h}{T_{v,0}}\right)^{\frac{9.8}{R_0\gamma_{v,g}}} = p_0\left(1 - \frac{h}{H}\right)^{\frac{9.8}{R_0\gamma_{v,g}}} \qquad (14.74)$$

where p_0 is the surface pressure. The density distribution within the polytropic atmosphere is easily found from the ideal gas law together with (14.73) and (14.74). The result is given by

$$\rho(h) = \rho(h = 0)\left(1 - \frac{h}{H}\right)^{\frac{9.8}{R_0\gamma_{v,g}} - 1} \tag{14.75}$$

A particularly interesting polytropic atmosphere refers to dry adiabatic stratification. Logarithmic differentiation of the potential temperature formula (12.7), holding θ constant and application of the hydrostatic equation gives

$$\boxed{-\left(\frac{d_g T}{dh}\right)_{\theta=\text{constant}} = \gamma_{a,g} = \frac{9.8}{c_{p,0}} = 0.98\,\text{K}/100\,\text{gpm}} \tag{14.76}$$

where the suffix a refers to adiabatic. This quantity is of great interest in atmospheric stability considerations. We shall take up this subject in more detail in the following sections.

14.7 Individual vertical temperature changes

In order to discuss the atmospheric stability of vertically displaced air parcels caused by solar heating of the ground or by some lifting process, we need to introduce the temperature change of the individual air parcel. In all considerations we shall assume that the vertical displacement takes place adiabatically accompanied either by reversible or irreversible processes. The dry adiabatic displacement follows as a special case. In order to make the analysis analytically tractable, we shall assume that the surrounding air is at rest and in hydrostatic equilibrium. Furthermore, we shall assume that the displacement occurs quasi-statically. By this we mean that the geometrical and the individual pressure change of the displaced air parcel are identical. Hence

$$dp = d_g p = -\frac{9.8p}{R_0 T_{v,g}}\,dh \tag{14.77}$$

where we have used the hydrostatic equation as assumed. All considerations refer to water clouds. From (12.16) and (12.40) we find for the reversible and irreversible processes

$$c_p\,dT - (R_0 m^0 + R_1 m^1)\frac{T}{p}\,dp = \begin{cases} -l_{21}\,dm^1 & \text{reversible} \\ -l_{21}m^0\,d\left(\frac{m^1}{m^0}\right) & \text{irreversible} \end{cases} \tag{14.78}$$

Since we are working with the filtered system we introduce the saturation mixing ratio (10.66) which is repeated for convenience

$$\frac{m^1}{m^0} = r^{21}(p, T) = \frac{R_0}{R_1} \frac{p^{21}(T)}{p - p^{21}(T)} \tag{14.79}$$

Logarithmic differentiation of this expression gives

$$\frac{dr^{21}}{r^{21}} = \frac{dp^{21}}{p^{21}} - \frac{d(p - p^{21})}{p - p^{21}} = \frac{p}{p - p^{21}} \left(\frac{dp^{21}}{p^{21}} - \frac{dp}{p} \right) \tag{14.80}$$

Using the definition of the saturation mixing ratio we find

$$\frac{p}{p - p^{21}} = 1 + \frac{p^{21}}{p - p^{21}} = 1 + \frac{r^{21}}{R_0/R_1} = 1 + \frac{R_1 m^1}{R_0 m^0} = \frac{R_0 m^0 + R_1 m^1}{R_0 m^0} \tag{14.81}$$

so that

$$\frac{dr^{21}}{r^{21}} = \frac{R_0 m^0 + R_1 m^1}{R_0 m^0} \left(\frac{d \ln p^{21}}{dT} dT - \frac{dp}{p} \right) \tag{14.82}$$

First let us consider the reversible system in (14.78) which is characterized by $dm^0 = 0$ so that $dm^1 + dm^2 = 0$. From logarithmic differentiation of (14.79) we obtain

$$dm^1 = m^1 \frac{dr^{21}}{r^{21}} = r^{21} \frac{R_0 m^0 + R_1 m^1}{R_0} \left(\frac{d \ln p^{21}}{dT} dT - \frac{dp}{p} \right) \tag{14.83}$$

where use has been made of the previous formula. Introducing the virtual temperature

$$T = \frac{R_0(m^0 + m^1)}{R_0 m^0 + R_1 m^1} T_v \tag{14.84}$$

as follows from (8.50) and (8.47), we finally get for the reversible process the relation

$$c_p \left(1 + r^{21} \frac{l_{21}}{c_p} (m^0 + m^1) \frac{T_v}{T} \frac{d \ln p^{21}}{dT} \right) dT - R_0 (m^0 + m^1) T_v \left(1 + \frac{l_{21} r^{21}}{R_0 T} \right) \frac{dp}{p} = 0 \tag{14.85}$$

For the irreversible process we obtain the same formula assuming that the liquid water leaves the system as soon as it forms. However, the exact agreement for the reversible and the irreversible process is formal only. For example, $m^0 + m^1 = 1$ for the irreversible process but $m^0 + m^1 = 1 - m^2$ for the reversible process. The numerical effect on dT/dt, however, is very small. The individual temperature change is found by substituting the quasi-static relation (14.77) into (14.85). We obtain

$$-\frac{dT}{dh} = \frac{9.8}{c_p} \left(\frac{(m^0 + m^1)\left(1 + \frac{l_{21} r^{21}}{R_0 T}\right)}{1 + (m^0 + m^1) r^{21} \frac{l_{21} T_v}{c_p T} \frac{d \ln p^{21}}{dT}} \right) \frac{T_v}{T_{v,g}} = \gamma \frac{T_v}{T_{v,g}} \tag{14.86a}$$

The temperature of the surrounding air is denoted by $T_{v,g}$, all remaining quantities refer to the individual air parcel. In general, the thermodynamic properties of the individual air parcel and the surrounding air will be different. If they coincide then we have $T_v = T_{v,g}$ and we have the autobarotropic situation. In this case the surrounding atmosphere is moist adiabatically stratified and the lapse rates are identical. Thus $\gamma = \gamma_g$ and we speak of the *moist adiabatic lapse rate*. In case of an ice cloud we must formally replace the sub- or superscript $k = 2$ by $k = 3$.

Sometimes it is of advantage to replace T in (14.86a) by the virtual temperature. This can be done with the help of (14.84). The exact form of the resulting equation is derived in the exercises. Presently we simply write

$$-\frac{dT_v}{dh} = \gamma_v \frac{T_v}{T_{v,g}} \tag{14.86b}$$

Finally, we need to mention two special cases dealing with completely dry air and with moist unsaturated air. From (12.16) we find for the moist unsaturated air $(m^2 = 0)$

$$c_p \frac{dT}{T} - R_m \frac{dp}{p} = 0 \quad \text{with} \quad c_p = c_{p,0} m^0 + c_{p,1} m^1 \tag{14.87}$$

where the suffix m denotes moist air. Substituting the quasi-static relation (14.77) for dp/p and observing that $R_m T = R_0 T_v$ we find an expression corresponding to (14.86a)

$$-\frac{dT}{dh} = \frac{9.8}{c_p} \frac{T_v}{T_{v,g}} = \gamma_m \frac{T_v}{T_{v,g}} \quad \text{with} \quad \gamma_m = \frac{9.8}{c_p} \tag{14.88}$$

This form also follows from (14.86a) by setting the terms involving l_{21} equal to zero and by observing that $m^0 + m^1 = 1$. In case of completely dry air equation (14.88) reduces to

$$-\frac{dT}{dh} = \frac{9.8}{c_{p,0}} \frac{T}{T_g} = \gamma_d \frac{T}{T_g} \quad \text{with} \quad \gamma_d = \frac{9.8}{c_{p,0}} \tag{14.89}$$

Inspection shows that the dry adiabatic displacement results in a slightly larger lapse rate than in case of moist air since $c_p > c_{p,0}$. For most practical situations the difference between these lapse rates is negligibly small and we may set $\gamma_m = \gamma_d$. Finally, in case of autobarotropy $(T = T_g)$ we obtain the dry adiabatic lapse rate $\gamma_d = -dT/dh = 0.98$ K/100 gpm.

14.8 Stability of the hydrostatic equilibrium

In this very important section we are going to derive at various levels of sophistication some stability criteria for the behavior of vertically displaced air parcels.

The following derivations are based on the fact that an air parcel will rise, sink or remain at the same level depending on whether the buoyancy force is greater than, less than, or equal to the downward force acting on the particle due to the acceleration of gravity.

14.8.1 Parcel-dynamic derivation of the hydrostatic stability criteria

We consider an individual air parcel which is displaced adiabatically and quasi-statically from an equilibrium level h_0 in the vertical direction. We assume that during this displacement the surrounding air remains at rest. While the displaced parcel and the surroundings in general differ in temperature and density, the pressure inside and outside of the air parcel are identical due to the quasi-static assumption. At the equilibrium level all thermodynamic variables of the air parcel and the surroundings coincide. In general the atmosphere will not be autobarotropic. This means that changes in virtual temperature and density experienced by the air parcel differ from the geometric change of the surroundings: $dT_v \neq d_g T_v$, $d\rho \neq d_g \rho$ but $dp = d_g p$. We may then conclude that in general the coefficients of piezotropy and barotropy will differ also,

$$\Gamma_p^\rho - \gamma_p^\rho \neq 0 \qquad (14.90)$$

The coefficient of barotropy is given by equation (14.65). With the help of the hydrostatic equation (14.61) and the ideal gas law we find for the coefficient of piezotropy by obvious steps

$$\gamma_p^\rho = \frac{d\rho}{dp} = \frac{d\rho}{d_g p} = \frac{1}{9.8} \frac{\rho}{\rho_g T_{v,g}} \left(\frac{9.8}{R_0} - \gamma_v \right) \quad \text{with} \quad \gamma_v = -\frac{T_{v,g}}{T_v} \frac{dT_v}{dh} \qquad (14.91)$$

The statement (14.90) is then equivalent to

$$9.8\rho_g \left(\frac{\Gamma_p^\rho}{\rho_g} - \frac{\gamma_p^\rho}{\rho} \right) = \frac{9.8\rho_g}{dp} \left(\frac{d_g\rho}{\rho_g} - \frac{d\rho}{\rho} \right) = \frac{1}{T_{v,g}} (\gamma_v - \gamma_{v,g}) \neq 0 \qquad (14.92)$$

with $dp = d_g p = -9.8\rho_g \, dh$. Due to the quasi-static assumption we have replaced T_v by $T_{v,g}$.

In order to interpret equation (14.92) consider Figure 14.3 showing the density changes experienced by the displaced air parcel and the geometric change of the surrounding air. Three different cases may occur. In case (1) the geometric change of the density exceeds the individual change. We find immediately that above and below the equilibrium level h the densities differ as shown in the figure. For example, above h_0 the density of the air parcel exceeds the geometric value causing the displaced air to return to the equilibrium level. Since the air parcel below the equilibrium level is lighter than the surrounding air it will also return to

$$h_0 + dh \qquad \boxed{\rho = \rho_0 - d\rho} \qquad \rho_g = \rho_0 - d_g\rho$$

$$\uparrow$$

$$h_0 \qquad \boxed{\rho = \rho_0} \qquad \rho_g = \rho_0$$

$$\downarrow$$

$$h_0 - dh \qquad \boxed{\rho = \rho_0 + d\rho} \qquad \rho_g = \rho_0 + d_g\rho$$

$$(1) \quad d\rho < d_g\rho \implies \left\{ \begin{array}{ll} h = h_0 + dh : & \rho > \rho_g \\ h = h_0 - dh : & \rho < \rho_g \end{array} \right\} \implies \gamma_v - \gamma_{v,g} > 0$$

$$(2) \quad d\rho = d_g\rho \implies \left\{ \begin{array}{ll} h = h_0 + dh : & \rho = \rho_g \\ h = h_0 - dh : & \rho = \rho_g \end{array} \right\} \implies \gamma_v - \gamma_{v,g} = 0$$

$$(3) \quad d\rho > d_g\rho \implies \left\{ \begin{array}{ll} h = h_0 + dh : & \rho < \rho_g \\ h = h_0 - dh : & \rho > \rho_g \end{array} \right\} \implies \gamma_v - \gamma_{v,g} < 0$$

Fig. 14.3 Stability of the hydrostatic equilibrium.

the original position. This situation is indicative of *atmospheric stability*. In case (2) the air parcel remains at rest at any position if there is no further cause for any displacement. This is the *indifferent equilibrium* between the air parcel and the surroundings. In case (3) the inequalities of the densities show that this is the *unstable case*. A slight displacement from the equilibrium level is sufficient for the air parcel to continue its motion either in upward or downward direction. From (14.92) we may summarize

$$\gamma_v - \gamma_{v,g} \left\{ \begin{array}{ll} > 0 & \text{stable equilibrium} \\ = 0 & \text{neutral equilibrium} \\ < 0 & \text{unstable equilibrium} \end{array} \right. \tag{14.93}$$

where the special case of neutral equilibrium has also been included in this formula.

An application of this formula is demonstrated in Figure 14.4 for a specific geometric temperature distribution. Inspection shows that region I is absolutely stable meaning that it is stable with respect to the moist and to the dry adiabat. Region II is indifferent or neutral with respect to the moist adiabat but stable with regard to the dry adiabat. Region III is absolutely unstable meaning that it is unstable with respect to the moist and the dry adiabatic lapse rate.

The general situation showing the regions of stability for the dry air and the moist saturated air are shown for the *Stüve diagram* in Figure 14.5. The same arrangement, of course, is valid for any other thermodynamic chart. Inspection shows that the unstable region is bounded by either the dry or the pseudoadiabatic lapse rate and by the lapse rate of the homogeneous atmosphere which cannot be exceeded. The

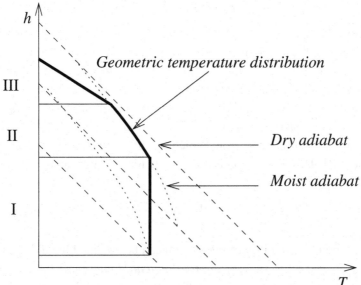

Fig. 14.4 Regions of stability for a particular example.

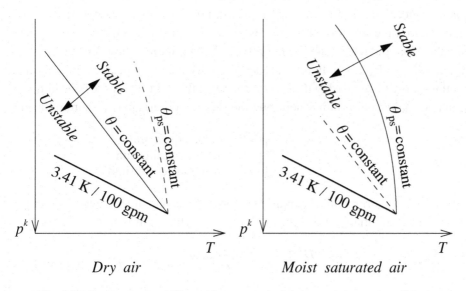

Fig. 14.5 Regions of stability with respect to the dry and the moist adiabat.

stable region is bounded either by the dry adiabatic or the pseudoadiabatic lapse rate. There is no theoretical limit for the most stable case.

14.8.2 Derivation of the stability criteria by energy variation

The derivation of the stability criteria (14.93) was based on the unrealistic assumption that the displaced air parcels did not disturb the surrounding air which remained completely at rest. In this section we wish to investigate if the stability criteria are still valid if they are derived by the method of energy variation. The basis of the derivation is the principle of energy conservation which applies to a system which is completely closed. For such a system we found an integral expression from (14.56) and (14.59) from which the hydrostatic equilibrium condition could be derived. This expression is repeated here for convenience

$$\frac{d}{dt}(E_{pot} + E_{int}) = \int_{V(t)} \rho\psi \, d\tau = 0 \quad \text{with} \quad \psi = \mathbf{v} \cdot (\nabla\phi + v\nabla p) \qquad (14.94)$$

Information about the stability of the system can be obtained by taking the second derivative with respect to time of the *total potential energy*. If the total potential energy has a minimum value, $d^2/dt^2(E_{pot} + E_{int}) > 0$, then every virtual displacement of the fluid must increase the total potential energy at the expense of the kinetic energy of any disturbance which might appear. Such a decrease in the kinetic energy of the fluid system is associated with a stable equilibrium. The virtual displacement must be distinguished from the true displacement which occurs in a certain time interval where forces and constraints may change. The opposite situation that the total potential energy has a maximum value is characterized by $d^2/dt^2(E_{pot} + E_{int}) < 0$. In this case a virtual displacement decreases the total potential energy by increasing the *kinetic energy* of the fluid. This process is indicative of an unstable equilibrium. Therefore, the stability criteria may be summarized by

$$\frac{d^2}{dt^2}(E_{pot} + E_{int}) \begin{cases} > 0 & \text{stable equilibrium} \\ = 0 & \text{neutral equilibrium} \\ < 0 & \text{unstable equilibrium} \end{cases} \qquad (14.95)$$

In order to interpret this inequality we form the second time derivative of (14.94) using the differentiation rules for fluid volumes. The result is shown next

$$\frac{d}{dt}\int_{V(t)} \rho\psi \, d\tau = \int_{V(t)} \frac{D(\rho\psi)}{Dt} \, d\tau = \int_{V(t)} \frac{\partial}{\partial t}(\rho\psi) \, d\tau + \int_{V(t)} \nabla \cdot (\rho\psi\mathbf{v}) \, d\tau$$

$$\text{with} \quad \int_{V(t)} \nabla \cdot (\rho\psi\mathbf{v}) \, d\tau = \int_{F(t)} \rho\psi\mathbf{v} \cdot d\mathbf{f} = 0$$

$$(14.96)$$

The integral containing the divergence term vanishes. This follows from the application of Gauss' divergence theorem since for the mechanically closed system the surface integral must be zero. Replacing the abbreviating symbol ψ as defined in

(14.94) we find

$$\frac{d^2}{dt^2}(E_{\text{pot}} + E_{\text{int}}) = \int_{V(t)} \frac{\partial}{\partial t}(\rho \mathbf{v}) \cdot (\nabla \phi + v \nabla p) \, d\tau$$

$$+ \int_{V(t)} \rho \mathbf{v} \cdot \frac{\partial}{\partial t}(\nabla \phi + v \nabla p) \, d\tau \tag{14.97}$$

The first term on the right-hand side represents the hydrostatic equilibrium condition (14.60) and, therefore, is zero. Since the geopotential is locally time independent, (14.97) after obvious mathematical steps, reduces to

$$\frac{d^2}{dt^2}(E_{\text{pot}} + E_{\text{int}}) = -\int_{V(t)} v \frac{\partial \rho}{\partial t} \mathbf{v} \cdot \nabla p \, d\tau + \int_{V(t)} \mathbf{v} \cdot \nabla \frac{\partial p}{\partial t} \, d\tau$$

$$= -\int_{V(t)} v \frac{\partial \rho}{\partial t} \mathbf{v} \cdot \nabla p \, d\tau + \int_{V(t)} \nabla \cdot \left(\mathbf{v} \frac{\partial p}{\partial t} \right) d\tau - \int_{V(t)} \frac{\partial p}{\partial t} \nabla \cdot \mathbf{v} \, d\tau \tag{14.98}$$

Analogously to (14.96) the second integral vanishes since the system is mechanically closed. In order to usefully interpret the remaining expression we assume that the energy transformation takes place dry adiabatically. This situation is described by (14.49) so that with the help of the continuity equation we find

$$v \frac{d\rho}{dt} = -\nabla \cdot \mathbf{v} = \frac{1}{\kappa p} \left(\frac{\partial p}{\partial t} + \mathbf{v} \cdot \nabla p \right) \tag{14.99}$$

From this formula we find the local time change of the density as needed by using the Euler expansion

$$v \frac{\partial \rho}{\partial t} = v \frac{d\rho}{dt} - v \mathbf{v} \cdot \nabla \rho = \frac{1}{\kappa p} \frac{\partial p}{\partial t} + \frac{1}{\kappa p} \mathbf{v} \cdot \nabla p - v \mathbf{v} \cdot \nabla \rho \tag{14.100}$$

Substitution of (14.99) and (14.100) into (14.98) gives

$$\frac{d^2}{dt^2}(E_{\text{pot}} + E_{\text{int}}) = \int_{V(t)} \mathbf{v} \cdot \left(v \nabla \rho - \frac{1}{\kappa p} \nabla p \right) \mathbf{v} \cdot \nabla p \, d\tau + \int_{V(t)} \frac{1}{\kappa p} \left(\frac{\partial p}{\partial t} \right)^2 d\tau \tag{14.101}$$

Since we have assumed dry adiabatic state changes it is best to introduce the potential temperature (12.7). Replacing the temperature by means of the ideal gas law in terms of density and pressure we find

$$\theta = \frac{p_0^{k_0} p^{1/\kappa}}{R_0 \rho}, \qquad k_0 = \frac{R_0}{c_{p,0}} = \frac{\kappa - 1}{\kappa}, \qquad \kappa = \frac{c_{p,0}}{c_{v,0}} \tag{14.102}$$

hence $\theta = \theta(p, \rho)$ and $\dfrac{1}{\theta} \nabla \theta = \dfrac{1}{\kappa p} \nabla p - v \nabla \rho$

showing that the potential temperature can be expressed as a function of p and ρ. Logarithmic differentiation of the potential temperature shows that the expression in parentheses in (14.101) may be replaced so that

$$\frac{d^2}{dt^2}(E_{\text{pot}} + E_{\text{int}}) = -\int_{V(t)} \left(\frac{1}{\theta}(\mathbf{v}\cdot\nabla\theta)(\mathbf{v}\cdot\nabla p)\right) d\tau + \int_{V(t)} \frac{1}{\kappa p}\left(\frac{\partial p}{\partial t}\right)^2 d\tau$$

(14.103)

Since we are dealing with hydrostatic equilibrium we may replace the pressure gradient by

$$\mathbf{v}\cdot\nabla p = -\rho\mathbf{v}\cdot\nabla\phi \qquad (14.104a)$$

Finally, we recall from (14.62e) that in case of hydrostatic equilibrium the potential temperature can be expressed solely as a function of the geopotential, i.e.

$$\theta = \theta(\phi) \implies \nabla\theta = \frac{d_g\theta}{d_g\phi}\nabla\phi \qquad (14.104b)$$

Using the information listed in (14.104) we obtain in place of (14.103)

$$\frac{d^2}{dt^2}(E_{\text{pot}} + E_{\text{int}}) = \int_{V(t)} \frac{\rho}{\theta}(\mathbf{v}\cdot\nabla\phi)^2 \frac{d_g\theta}{d_g\phi} d\tau + \int_{V(t)} \frac{1}{\kappa p}\left(\frac{\partial p}{\partial t}\right)^2 d\tau \qquad (14.105)$$

The second integral on the right-hand side involving the pressure tendency is positive definite. A comparison of the magnitudes of the two integrals shows that the second integral may be ignored. In the exercises it will be shown that this term represents the influence of the *adiabatic compressibility* of the air on the stability criteria. The sign of the remaining integral is now solely determined by the geometric change of the potential temperature with the geopotential. From (14.95) then follows that

$$\frac{d_g\theta}{d_g\phi} \begin{cases} > 0 & \text{stable equilibrium} \\ = 0 & \text{neutral equilibrium} \\ < 0 & \text{unstable equilibrium} \end{cases} \qquad (14.106)$$

It will be seen that the neglected term has a stabilizing tendency thus slightly changing the limits of the neutral equilibrium.

For comparison with the parcel dynamic method we need to give (14.106) a form similar to (14.93). Logarithmic differentiation of the potential temperature in its usual form (12.7) with respect to the geopotential, together with the hydrostatic approximation (14.61b), after a few elementary steps gives

$$\frac{1}{\theta}\frac{d_g\theta}{d\phi} = \frac{1}{T}\frac{d_gT}{d\phi} - \frac{R_0}{c_{p,0}}\frac{1}{p}\frac{d_gp}{g\phi} = -\frac{1}{9.8T}\gamma_g + \frac{1}{c_{p,0}T} \qquad (14.107)$$

from which we obtain

$$\frac{d_g\theta}{d\phi} = \frac{1}{9.8}\frac{\theta}{T}(\gamma_d - \gamma_g) \tag{14.108}$$

and finally

$$\gamma_d - \gamma_g \begin{cases} > 0 & \text{stable equilibrium} \\ = 0 & \text{neutral equilibrium} \\ < 0 & \text{unstable equilibrium} \end{cases} \tag{14.109}$$

The only difference is that (14.109) is valid only for dry adiabatic displacements whereas (14.93) also applies to moist air.

14.8.3 Vertical stability vibrations of isolated air parcels

Let us consider an isolated parcel of air which is given a vertical impetus while the surrounding atmosphere is assumed to be completely at rest. For this unrealistic situation, using the parcel dynamic method, we are able to set up an approximate equation of motion for the displaced particle. The solution to this equation gives information on the stability of the atmosphere. Neglecting the Coriolis effect and friction, the vertical component of the equation of motion due to the quasi-static condition may be written as

$$\frac{dw}{dt} = -g - v\frac{\partial p}{\partial z} = -g + g\frac{\rho_g}{\rho} = g\frac{\rho_g - \rho}{\rho} = g\frac{T_v - T_{v,g}}{T_{v,g}} = A \gtreqless 0$$

$$\text{with}\quad A \begin{cases} > 0 & \text{upward acceleration} \\ = 0 & \text{no acceleration} \\ < 0 & \text{downward acceleration} \end{cases}$$

$$\tag{14.110}$$

The quantity A is known as the *vertical acceleration* or the *buoyancy*. As usual, positive and negative A represent upward and downward accelerations, respectively. If the parcel is at rest at z_0 or at the geopotential height h_0, we may write for the vertical acceleration

$$\frac{dw}{dt} = \frac{d^2}{dt^2}(z - z_0) = \frac{9.8}{g}\frac{d^2}{dt^2}(h - h_0) \tag{14.111}$$

so that equation (14.110) can be rewritten as

$$\frac{d^2}{dt^2}(h - h_0) = \frac{g^2}{9.8}\frac{\rho_g - \rho}{\rho} = \frac{g^2}{9.8}\frac{T_v - T_{v,g}}{T_{v,g}} \tag{14.112}$$

Substituting equation (14.86b) into (14.112), the equation of motion assumes the form

$$\frac{d^2}{dt^2}(h - h_0) + \omega^2(h - h_0) = 0 \tag{14.113}$$

with

$$\omega^2 = \frac{g^2}{9.8(h - h_0)} \frac{1}{\gamma_v} \left[\gamma_v - \left(-\frac{dT_v}{dh} \right) \right] \tag{14.114}$$

Equation (14.113) would be a well-known constant coefficient differential equation if ω^2 is a constant. Inspection shows that this is not the case since ω^2 depends on $h - h_0$. However, by introducing a polytropic atmosphere for the surrounding air, ω^2 will be at least approximately constant as we will now show. Integration of (14.86b) from h_0 to h, using a linear variation of $T_{v,g} = T_{v,0} - \gamma_{v,g}(h - h_0)$ with $T_{v,0} = T_v(h_0) = T_{v,g}(h_0)$ assuming the ratio of the lapse rates to be a constant, we obtain after a few easy steps

$$T_v = T_{v,0} \left(\frac{T_{v,g}}{T_{v,0}} \right)^{\gamma_v/\gamma_{v,g}} \tag{14.115}$$

Thus the lapse rate of the virtual temperature can now be approximated by

$$-\frac{dT_v}{dh} \approx \gamma_v \left(1 - \frac{\gamma_v - \gamma_{v,g}}{T_{v,0}}(h - h_0) \right) \tag{14.116}$$

so that (14.114) assumes the simplified form

$$\boxed{\omega^2 = \frac{g^2}{9.8} \frac{\gamma_v - \gamma_{v,g}}{T_{v,0}} \approx \text{constant}} \tag{14.117}$$

The quantity ω is known as the *Brunt–Vaisala frequency*.

Obviously, the solution of (14.113) depends on the sign of ω^2. In case that $\omega^2 > 0$ the solution is a pure oscillation about h_0 indicating *stable atmospheric stratification*, that is

$$\omega^2 > 0, \qquad \gamma_v > \gamma_{v,g}$$

$$h - h_0 = C_1 \cos \omega t + C_2 \sin \omega t, \qquad \omega = \frac{2\pi}{\tau} = \sqrt{\frac{g^2}{9.8 T_{v,0}}(\gamma_v - \gamma_{v,g})} \tag{14.118a}$$

The integration constants C_1, C_2 can be found from the initial conditions. The period of the oscillation τ is easily determined from the Brunt–Vaisala frequency which depends mainly on the difference of the lapse rates.

In case that $\omega^2 < 0$ we get the exponential solution indicating *unstable stratification*

$$\omega^2 < 0, \qquad \gamma_v < \gamma_{v,g}$$

$$h - h_0 = C_1 \exp(\omega t) + C_2 \exp(-\omega t) \tag{14.118b}$$

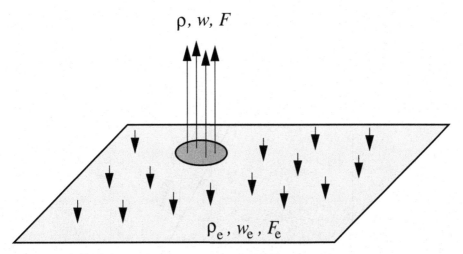

ρ, w, F

ρ_e, w_e, F_e

Fig. 14.6 Qualitative representation of the compensated motion through the reference surface.

As before the integration constants may be determined from the initial conditions. This solution refers to an unstable atmospheric stratification since the displacement of an air parcel from its resting position at h_0 occurs exponentially in time.

Finally, in case that $\omega^2 = 0$ we have the neutral or the *indifferent solution*

$$\omega^2 = 0, \qquad \gamma_v = \gamma_{v,g}$$

$$h - h_0 = C_1 t + C_2 \tag{14.118c}$$

In this case the atmosphere is neutrally stratified indicating that a displaced parcel of air is leaving the original position h_0 with a constant velocity corresponding to the initial momentum imparted to it.

14.8.4 The slice method

In Sections 14.8.1 and 14.8.3 we have assumed that the displaced air parcel does not disturb the surrounding atmosphere, which is to remain completely at rest. In reality, the vertical motion of a displaced air parcel requires a compensating vertical motion by the surrounding atmosphere. The effect of the compensating motion is treated in the so-called slice method devised by Bjerkness (1938) which will be discussed now. This method assumes that a vertical current of saturated air of density ρ and of a relatively small cross-section F is rising with a relatively large vertical velocity w. This motion is compensated by descending unsaturated air of the environment of density ρ_e with relatively small vertical velocity w_e which is spread over the relatively large area F_e as shown in Figure 14.6.

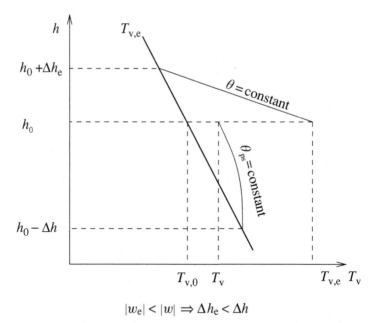

$$|w_e| < |w| \Rightarrow \Delta h_e < \Delta h$$

Fig. 14.7 Determination of temperatures for equations (14.120) and (14.121).

Assuming that on the average the upward and downward mass flux through the reference plane is completely balanced we may write

$$\rho\,|w|\,F = \rho_e\,|w_e|\,F_e \qquad (14.119)$$

In order to determine the stability criteria we must set up the equation for the vertical motion of the air parcel. Figure 14.7 will be used to determine the required virtual temperatures at the reference height.

The saturated air assumed to originate at the height $h_0 - \Delta h$ is rising moist adiabatically. At height h_0 the virtual temperature is given by

$$T_v(h_0) = T_{v,0} + \gamma_{v,g}\Delta h - \gamma_v \frac{T_v}{T_{v,g}}\Delta h \approx T_{v,0} - \Delta h(\gamma_v - \gamma_{v,g}) \qquad (14.120)$$

since the ratio $T_v/T_{v,g} \approx 1$. In contrast, the moist but unsaturated environmental air is assumed to descend from $h_0 + \Delta h_e$ to the reference height. In good approximation the descending air moves dry adiabatically so that its temperature at h_0 is given by

$$T_{v,e}(h_0) = T_{v,0} - \gamma_{v,g}\Delta h_e + \gamma_d\Delta h_e \approx T_{v,0} + \Delta h_e(\gamma_d - \gamma_{v,g}) \qquad (14.121)$$

The equation of motion in the form (14.112) must now be modified to handle the present situation. Therefore, we use the replacements

$$h - h_0 \longrightarrow \Delta h, \qquad T_{v,g} \longrightarrow T_{v,e} \qquad (14.122)$$

and obtain

$$\frac{d^2 \Delta h}{dt^2} = \frac{g^2}{9.8} \frac{T_v - T_{v,e}}{T_{v,e}} \approx \frac{g^2}{9.8} \frac{T_v - T_{v,e}}{\overline{T}_{v,e}} \tag{14.123}$$

where the temperature in the denominator has been replaced by an average value in the domain of interest ranging from $h_0 - \Delta h$ to $h_0 + \Delta h_e$. Substitution of (14.120) and (14.121) into (14.123) and using the approximation

$$\frac{\Delta h_e / \Delta t}{\Delta h / \Delta t,} \approx \frac{|w_e|}{|w|} = \frac{\rho F}{\rho_e F_e} \tag{14.124}$$

with (14.119) gives the equation of the rising saturated air

$$\frac{d^2 \Delta h}{dt^2} + \frac{g^2}{9.8 \overline{T}_{v,e}} (\gamma_v - \gamma_{v,g}) \left[1 + \frac{\rho F}{\rho_e F_e} \frac{\gamma_d - \gamma_{v,g}}{\gamma_v - \gamma_{v,g}} \right] \Delta h = 0 \tag{14.125a}$$

or

$$\frac{d^2 \Delta h}{dt^2} + \omega^{*2} \Delta h = 0 \quad \text{with}$$

$$\omega^{*2} = \frac{g}{\overline{T}_{v,e}} (\gamma_v - \gamma_{v,g}) \left[1 + \frac{\rho F}{\rho_e F_e} \frac{\gamma_d - \gamma_{v,g}}{\gamma_v - \gamma_{v,g}} \right] \tag{14.125b}$$

The symbol $*$ has been added to ω^2 to show that now compensating motion has been taken into account. By comparing ω^{*2} with the corresponding expression (14.117) of the previous section it is seen that the fraction in the brackets represents the modification due to the compensating air motion. In order to obtain a simple analytic solution we must assume that ω^{*2} is a constant which is realized only approximately. The type of solution depends on the sign of ω^{*2}. The case $\omega^{*2} > 0$ yields the vibrational solution representing the stable case. If $\omega^{*2} < 0$ or $\omega^{*2} = 0$ we obtain the unstable and the neutral situations.

There are two possible situations to render $\omega^{*2} > 0$ which are listed in

$$\omega^{*2} > 0: \quad \begin{array}{ll} (\alpha) & \gamma_{v,g} < \gamma_v < \gamma_d \\[2mm] (\beta) & \gamma_v < \gamma_{v,g} < \gamma_d \quad \text{and} \quad 1 + \frac{\rho F}{\rho_e F_e} \frac{\gamma_d - \gamma_{v,g}}{\gamma_v - \gamma_{v,g}} < 0 \end{array} \tag{14.126a}$$

If the inequality listed under (α) is obeyed then the expression in brackets is positive also. Somewhat more complicated is the case (β) where the difference in lapse rates in parentheses is a negative quantity so that the expression in the brackets must also be negative. The unstable case may occur if

$$\omega^{*2} < 0: \quad \gamma_{v,g} > \gamma_v \quad \text{and} \quad 1 + \frac{\rho F}{\rho_e F_e} \frac{\gamma_d - \gamma_{v,g}}{\gamma_v - \gamma_{v,g}} > 0 \tag{14.126b}$$

The limiting case $\omega^{*2} = 0$ describes the indifferent or neutral equilibrium.

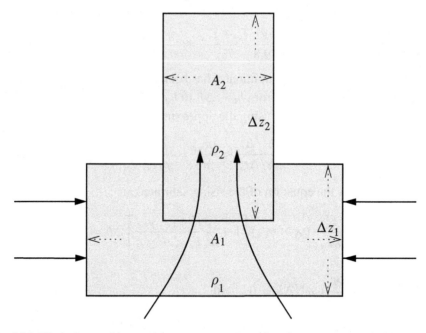

Fig. 14.8 Vertical stretching and convergence of an air column near the earth's surface.

14.8.5 Effects of vertical motion on the stability of an air column

Let us consider an air column defined by its thickness Δz, the average density ρ and the area A of its projection onto the horizontal plane. We wish to estimate the stability changes that might occur if the air column is transformed. Let the subscripts 1 and 2, respectively, refer to the air column in the initial and final state. Assuming mass conservation we obtain

$$M = \rho_1 A_1 \Delta z_1 = \rho_2 A_2 \Delta z_2 \quad \text{or} \quad \left(\frac{\rho_1}{\rho_2}\right)\left(\frac{\Delta z_1}{\Delta z_2}\right) = \frac{A_2}{A_1} \qquad (14.127)$$

Figure 14.8 shows a situation where the air column stretches due to the influence of convergence. By reversing the direction of the arrows and interchanging the subscripts the figure refers to vertical shrinking and divergence.

We will now discuss a few artificial cases that give some insight into the ways in which the stability of an air column may change. First we simply ignore any divergence (convergence) effects so that $A_1 = A_2$. If the layer ascends then $\rho_2 < \rho_1$ so that stretching takes place since $\Delta z_2 > \Delta z_1$. Consider the simple case of a constant lapse rate and dry adiabatic lifting as shown on the Skew T-Log$p$ diagram, Figure 14.9. In the present situation the stability parameter is $\Delta\theta/\Delta z$. Here $\Delta\theta = $ constant while the height of the column increases from Δz_1 to Δz_2 assuming no change in the cross-section area. The stretching of the column is

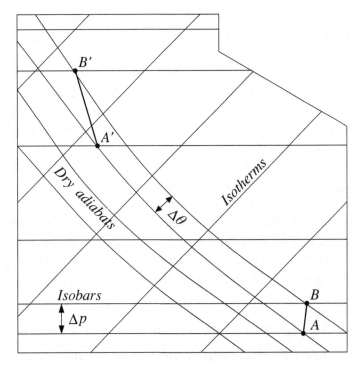

Fig. 14.9 Lapse rate changes due to a dry adiabatic displacement, schematic view.

due to the decrease of pressure with height. Therefore, the stability of the column decreases due to stretching and the lapse rate approaches the dry adiabatic value. Had the layer been initially unstable, the lifting process would have decreased the instability. The opposite is true in case of a dry adiabatic descent thus increasing the stability of an already stable layer but increasing the instability of an unstable layer. Summarizing: lifting causes the lapse rate to approach the dry adiabatic lapse rate while sinking causes the lapse rate to depart from it. The same reasoning applies to the moist adiabatic case if the entire displacement takes place along the moist adiabat.

Now we consider the special case that convergence takes place without vertical motion so that $\rho_1 = \rho_2$ and $\Delta z_1/\Delta z_2 = A_2/A_1$ with $A_1 > A_2$ so that stretching occurs with $\Delta z_2 > \Delta z_1$. Therefore, the stability is changed in the same sense as in the case of adiabatic lifting without convergence. If divergence occurs with $A_2 > A_1$ then the thickness of the layer decreases. Hence, the lapse rate changes in the same sense as a descent without divergence. Therefore, horizontal convergence (divergence) tends to cause an additional vertical stretching (shrinking) of a rising (sinking) air column. This fact has a practical significance in the lower atmosphere due to the constraining atmospheric boundary at the earth's surface where horizontal convergence (divergence) must always accompany ascent (descent). It should be noted that subsiding air motion often produces very stable layers known as *subsidence inversions*.

In the upper layers of the atmosphere any combination of ρ, Δz and A may occur which satisfy the continuity equation (14.127), but this makes it more difficult to assess stability changes. More information on this subject may be found in other textbooks. Our reference goes to the Weather Service Manual *Use of the Skew T-Logp Diagram in Weather Service and Forecasting*.

One final remark. Sometimes the concept of *potential stability*, also called *convective stability*, is mentioned in the literature. This refers to stability of a layer if the entire layer has been lifted to saturation. If the layer is stable after lifting to saturation, independent of the initial lapse rate, it is called potentially stable. Should the layer be unstable after lifting it is termed potentially unstable. More details can be found in the same manual or, for example, in the textbook *Dynamical and Physical Meteorology* by Haltiner and Martin.

14.9 Atmospheric energetics of the hydrostatic equilibrium

14.9.1 Buoyancy energy and atmospheric stability

The task ahead is to find expressions for energy changes experienced by a vertically displaced air parcel. From such expressions we will then derive information about the stability of the atmosphere. It is assumed that the air parcel is closed with respect to mass and heat transfer and that the displacement occurs quasi-statically. The mathematical analysis then proceeds from a simplified form of the first law of thermodynamics (3.36) not only assuming adiabatic conditions but also frictionless transformations

$$\frac{dh}{dt} - v\frac{dp}{dt} = 0 \tag{14.128}$$

where h denotes the specific enthalpy of the system. Using the quasi-static condition

$$dp = d_g p = -\rho_g \, d\phi \tag{14.129}$$

we obtain

$$dh + d\phi + B d\phi = 0 \quad \text{with} \quad B = \frac{\rho_g - \rho}{\rho} \tag{14.130}$$

If B is multiplied by g we have the acceleration A due to the buoyancy of the air parcel. Integrating this expression immediately yields

$$h + \phi + \int_{\phi_N}^{\phi} \frac{\rho_g - \rho}{\rho} \, d\phi' = h_N + \phi_N \tag{14.131}$$

where the subscript $_N$ refers to the original position of the parcel. Using (14.110) the integral term can be written as

$$\int_{\phi_N}^{\phi} \frac{\rho_g - \rho}{\rho} \, d\phi' = \int_{z_N}^{z} A \, dz' = \int_{z_N}^{z} \frac{dw}{dt} \, dz' \qquad (14.132)$$

representing the *specific buoyancy energy* e_b involved in the displacement of the air parcel. Since the density of the displaced parcel generally differs from the density distribution of the surroundings, e_b may be either positive, negative and sometimes even zero

$$e_b = \int_{\phi_N}^{\phi} \frac{\rho_g - \rho}{\rho} \, d\phi' \begin{cases} > 0 & \text{positive buoyancy energy} \\ = 0 & \text{no buoyancy energy} \\ < 0 & \text{negative buoyancy energy} \end{cases} \qquad (14.133)$$

Negative buoyancy energy exists whenever the atmosphere is characterized by stable stratification whereas in an unstable atmosphere buoyancy energy is released. Since the density of the atmosphere is not an observed variable of state it is profitable to introduce the temperature. With the help of the ideal gas law, again using the quasi-static assumption, we find for the buoyancy energy

$$e_b = \int_{p}^{p_N} \frac{T_v - T_{v,g}}{T_{v,g}\rho_g} \, dp' = R_0 \int_{p}^{p_N} (T_v - T_{v,g}) \, d \ln p'$$
$$= \int_{z_N}^{z} \frac{dw}{dz'} \frac{dz'}{dt} \, dz' = \int_{z_N}^{z} \frac{d}{dz'} \left(\frac{w^2}{2} \right) dz' = \frac{w^2}{2} - \frac{w_N^2}{2} \qquad (14.134)$$

permitting us to estimate the kinetic energy per unit mass as well as the vertical velocity. Usually it is assumed that the vertical velocity is zero at the level from which the displacement takes place.

Formula (14.134) forms the basis for a somewhat crude stability evaluation of the entire atmosphere. This will be explained next with the help of Refsdahl's *emagram* which was already discussed in the previous chapter. Consider the atmospheric sounding of the virtual temperature as shown in Figure 14.10.

In order to determine the stability of the atmosphere we follow an unsaturated air parcel rising from the ground to the equilibrium level, EL. In the language of thermodynamics the air parcel at the ground is a homogeneous closed system of moist unsaturated air which is forced to rise dry adiabatically. Forced rising occurs, for example, if air crosses a mountain range. The question arises at which level saturation and cloud formation occurs. This level is known as the *lifting condensation level*, LCL. The rising system remains unsaturated up to the LCL thus conserving the observed ground level mixing ratio. This mixing ratio is equivalent to the saturation mixing ratio specified by the ground pressure and the dew point

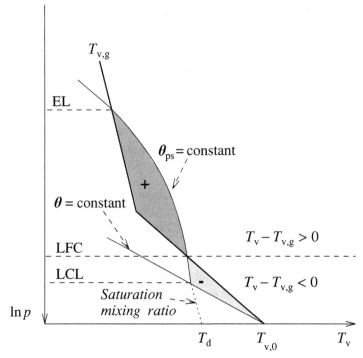

Fig. 14.10 Buoyancy energy or latent instability.

temperature T_d. Therefore, the LCL is found at the point where the saturation mixing ratio line proceeding from T_d intersects the dry adiabat starting from $T_{v,0}$ as shown in Figure 14.10. Any further forced upward displacement now proceeds along the pseudoadiabat to the *level of free convection*, LFC where the temperature of the parcel and of the sounding coincide. Between its initial position at the ground and the LFC the parcel is colder than the atmosphere so that the density of the parcel exceeds the density of the surrounding air. This corresponds to negative buoyancy energy or to the energy that must be supplied to the air parcel to be lifted from the ground to the level of free convection. For this reason the area between the two soundings is given the negative sign. From now on the parcel is less dense than the surrounding air so it rises to the equilibrium level EL without any further energy supply by the lifting process. This corresponds to a positive buoyancy energy or a positive area. The sounding is considered stable when no positive area exists. Whenever the negative area exceeds the positive area the atmosphere possesses *pseudo-latent instability*. Finally, when the positive area exceeds the negative area the atmosphere is classified as possessing real *latent instability*.

   This type of classification is somewhat insufficient to describe the stability of the atmosphere since it is solely based on the moisture content of the parcel at the surface and the temperature distribution of the atmosphere. In reality some mixing of the parcel with the surrounding atmosphere will take place. This mixing or entrainment effect is quite difficult to assess.

Another way to classify the stability of the atmosphere is by means of a *stability index*. Various proposals for the computation of such an index have been made. A particularly simple yet quite useful index is computed by lifting an air parcel from some fixed level such as the 850 hPa level dry adiabatically to the LCL and then moist adiabatically to the 500 hPa level. The temperature difference at this level taken as the observed temperature minus the temperature of the parcel is called the stability index. The stability increases with increasing value of the index. From numerous observations it has been found, for example, that an index value of 3 or less is associated with showers, a value of −3 or less with severe thunderstorms. Further information on this particular index is given by Showalter (1953).

14.9.2 Energy transformation

We begin our discussion by estimating the potential energy contained in an air column of unit cross-section. In a similar way the internal energy of the air column can be estimated. The sum of these we have called the total potential energy. For a specific situation we will then estimate the generation of kinetic energy from a reduction of the total potential energy due to a decrease of the mean temperature of such an air column. Finally, we calculate the mean wind speed from the kinetic energy. In order to keep the calculation as simple as possible we assume that the air is completely dry.

According to (14.53) the potential energy of an air column of unit cross-section is given by

$$E^*_{pot} = \frac{1}{F} \int_V \rho\phi \, d\tau = \int_0^\infty \rho g z \, dz = -\int_{p_0}^0 z \, dp = -\int_{p_0}^0 d(pz) + \int_0^\infty p \, dz$$

$$(14.135)$$

The first integral on the right-hand side of this equation vanishes due to the chosen limits $(0, p_0)$ of the integral. With the help of the ideal gas law and the hydrostatic equation we may rewrite this expression as

$$E^*_{pot} = \int_0^\infty p \, dz = R_0 \int_0^\infty \rho T \, dz = R_0 \int_0^{p_0} \frac{T}{g} \, dp \qquad (14.136)$$

This equation can be easily evaluated by assuming a model temperature distribution or simply by assigning an average temperature $\overline{T}$ for the entire air column. For any reasonable value of $\overline{T}$ we obtain $E^*_{pot} \approx 10^9$ J m^{-2} for an air column extending vertically from the ground to the top of the atmosphere.

The internal energy of an air column of unit cross-section is given by

$$E^*_{int} = \frac{1}{F}\int_V \rho e\, d\tau = \int_0^\infty \rho e\, dz = \int_0^{p_0} \frac{e}{g}\, dp \qquad (14.137)$$

Introducing the internal energy from (8.81) we obtain

$$E^*_{int} = \int_0^{p_0} \frac{c_v}{g}(T - T_0)\, dp + e(T_0)\int_0^{p_0} \frac{1}{g}\, dp = \int_0^{p_0} \frac{c_v}{R_0}\frac{R_0 T}{g}\, dp + \text{constant} \qquad (14.138)$$

For convenience we have left off subscripts and the superscript since we are dealing with dry air only. While the potential energy of an air column could be computed without difficulty, it is not possible to evaluate the internal energy due to the unspecified reference value $e(T_0)$. For a rough estimate we select $T_0 = 0$ K. In this case $e(T_0)$ vanishes and so does the constant.

In general, the internal and the potential energies of an air column are related by

$$\boxed{E^*_{int} = \frac{c_v}{R_0}E^*_{pot} + \text{constant}} \qquad (14.139)$$

as follows from a comparison of (14.136) and (14.138). We recall that the total potential energy was defined as the sum of the potential and the internal energy so that

$$E^*_{pot} + E^*_{int} = \int_0^{p_0} \frac{c_p}{g}(T - T_0)\, dp + \int_0^{p_0} \frac{R_0}{g}T_0\, dp + e(T_0)\int_0^{p_0} \frac{1}{g}\, dp \qquad (14.140)$$

We will now show that this expression is equivalent to the enthalpy of an air column of unit cross-section. In analogy to (14.137) we may write

$$H^* = \frac{1}{F}\int_V \rho h\, d\tau = \int_0^\infty \rho h\, dz = \int_0^{p_0} \frac{h}{g}\, dp \qquad (14.141)$$

with $\quad h = c_p(T - T_0) + h(T_0) = c_p(T - T_0) + e(T_0) + R_0 T_0$

where the *enthalpy h* is taken from (8.83). Substitution of h gives

$$H^* = \int_0^{p_0} \frac{c_p}{g}(T - T_0)\, dp + \int_0^{p_0} \frac{R_0 T_0}{g}\, dp + e(T_0)\int_0^{p_0} \frac{1}{g}\, dp = E^*_{pot} + E^*_{int} \qquad (14.142)$$

showing that the *total potential energy* and the enthalpy are equivalent. For the special case that the reference temperature T_0 is taken as 0 K we find from (14.139) with $e(T_0) = 0$

$$\frac{E^*_{pot}}{E^*_{int}} = \frac{R_0}{c_v} \approx 0.4$$

$$E^*_{pot} = 10^9\,\mathrm{J\,m^{-2}}, \qquad E^*_{int} \approx 2.5 E^*_{pot}, \qquad H^* \approx 3.5 E^*_{pot} \qquad (14.143)$$

$$E^*_{kin} = \frac{1}{F}\int_V \rho \frac{v^2}{2}\, d\tau = \int_0^\infty \rho \frac{v^2}{2}\, dz = \int_0^{p_0} \frac{v^2}{2g}\, dp$$

showing that the enthalpy or the total potential energy is 3.5 times as large as the potential energy itself.

The formula for the *kinetic energy* of an air column per unit cross-section has been included in (14.143). This expression will be needed to estimate the generation of kinetic energy due to an assumed reduction of the mean temperature

$$\overline{T}_p = \frac{1}{p_0} \int_0^{p_0} T \, dp \tag{14.144}$$

of an air column. For an air volume which is closed with respect to mass and energy transfer, the energy conservation law is given by

$$E^*_{pot} + E^*_{int} + E^*_{kin} = \text{constant} \tag{14.145}$$

The generation of kinetic energy due to the change of the mean temperature can be estimated from (14.140) with $T_0 = 0$ as

$$
\begin{aligned}
\Delta E^*_{kin} &= -\frac{d}{d\overline{T}_p} \left(E^*_{pot} + E^*_{int} \right) \Delta \overline{T}_p = -\frac{d}{d\overline{T}_p} \left(\frac{c_p}{g} \int_0^{p_0} T \, dp \right) \Delta \overline{T}_p \\
&= -\frac{d}{d\overline{T}_p} \left(\frac{c_p}{g} p_0 \overline{T}_p \right) \Delta \overline{T}_p = -\frac{c_p p_0}{g} \Delta \overline{T}_p
\end{aligned} \tag{14.146}
$$

We observe that p_0/g is the total mass M contained in an air column of unit cross-section so that the generated kinetic energy for the nonturbulent atmosphere may be estimated from

$$\Delta E^*_{kin} = M \Delta \frac{\overline{v}^2}{2} = -M c_p \Delta \overline{T}_p$$

$$\Delta \frac{\overline{v}^2}{2} = -c_p \Delta \overline{T}_p \approx 10^3 \, \text{m}^2 \, \text{s}^{-2} \tag{14.147}$$

$$\overline{v} \approx 45 \, \text{m s}^{-1}$$

If the atmosphere is at rest initially, for an assumed decrease of the mean temperature of the air of only 1 K we find that a substantial value of the kinetic energy has been generated. From this we calculate an unrealistically high average velocity of 45 m s^{-1}. This example implies that only a small part of the total potential energy or the enthalpy is available for the generation of kinetic energy. This part is known as the *available potential energy*. Energy transformations of this type were first studied by Margules (1905) whose paper is still worth reading even today.

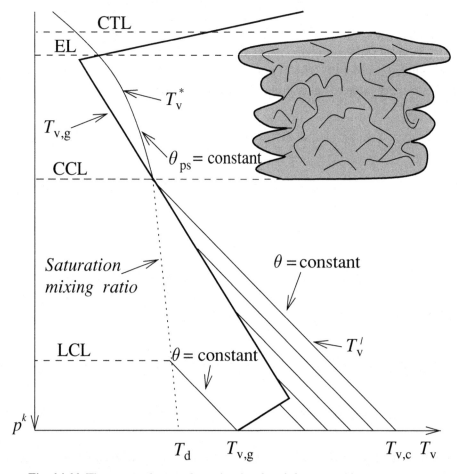

Fig. 14.11 The convective condensation level and the convective temperature.

14.9.3 The convective condensation level

It is well known that deep ground inversions form in clear calm nights as shown in
Figure 14.11. To give a little variety in the choice of the thermodynamic charts we
have used the frame of the *Stüve diagram*. The procedure to be described, however,
applies to any other thermodynamic chart as well. The temperature distribution
$T_{v,g}(p)$ of the atmosphere shows two inversions. The ground inversion forms as
the result of the nocturnal cooling of the ground. After sunrise effective heating of
the ground due to solar radiation may take place so that the resulting convection
will eliminate the ground inversion. Directly above the ground a relatively thin
extremely unstable air layer forms (not shown) which is characterized by a strong
superadiabatic lapse rate. This unstable temperature distribution initiates turbulent
mixing so that a mixed layer of constant potential temperature is formed. Continued

solar heating of the ground inceases the depth of the mixed layer so that air parcels rise to the *convective condensation level*, CCL where cloud formation takes place. The dry adiabat between the CCL and the ground defines the *convective temperature* $T_{v,c}$.

Graphically, see Figure 14.11, the CCL is found at the point where the saturation mixing ratio line representing the dew point at the ground crosses the geometric temperature distribution. After saturation at the CCL the air parcel rises along the pseudoadiabat due to the buoyancy energy represented by the positive area until it crosses the sounding at the equilibrium level EL. Here the momentum the parcel has gained will carry it to the cloud top level CTL some distance above the EL.

The area defined by the dry adiabat going from $T_{v,c}$ to the CCL and the geometric temperature distribution $T_{v,g}$ at sunrise represents the solar energy required to destroy the stable layer between the ground and the CCL. This energy can be calculated from the first law of thermodynamics assuming isobaric heating. For an air column of unit cross-section the required energy E_c is given by

$$E_c = \int c_p(T'_v - T_{v,g}) \, dM = \int_0^{z_{CCL}} c_p \rho (T'_v - T_{v,g}) \, dz = \int_{p_{CCL}}^{p(z=0)} \frac{c_p}{g}(T'_v - T_{v,g}) \, dp$$

$$(14.148)$$

Knowing this amount of energy for a particular day and location we can estimate from measurements of the solar radiation or from a radiation model the time of the day when this energy was supplied to the ground. This time roughly corresponds to the time at which convective clouds will form. It should be clear that this method can give a rough estimate only since frictional effects as well as entrainment have not been included in this discussion.

14.10 Problems

14.1: Verify equation (14.68b).

14.2: Consider the vertical distance between the 1000 hPa and the 500 hPa surfaces. If the barometric mean temperature changes by 2 K, find the corresponding change in gpm of the relative topography 500 hPa over 1000 hPa.

14.3: The *polytropic exponent n* is defined by

$$n = \frac{9.8}{R_0 \gamma_{g,v}} - 1$$

The scale height of pressure for the temperature $T_{v,0}$ is denoted by $H_{p,0}$.

(a) Show that

$$H = H_{p,0}(n + 1), \qquad p = p_0 \left(1 - \frac{h}{H}\right)^{n+1}$$

$$T_v = T_{v,0} \left(\frac{p}{p_0}\right)^{1/(n+1)}, \qquad h = H\left[1 - \left(\frac{p}{p_0}\right)^{1/(n+1)}\right]$$

$$\rho = \rho_0 \left(1 - \frac{h}{H}\right)^n, \qquad \Gamma_p^\rho = \frac{\gamma_{g,v}}{9.8T_v} n$$

(b) Show that for a homogeneous atmosphere the temperature and pressure decrease linearly with height. Find the height of the homogeneous atmosphere with $T_{v,0} = 273$ K.

14.4: Use the definition of the polytropic exponent $n = 9.8/(R_0\gamma_{g,v}) - 1$ to show that in a dry adiabatically stratified atmosphere the pressure and density distributions follow power laws. State the proper equations by assuming hydrostatic conditions.

14.5: Show that the integral term in equation (14.105) involving $(\partial p/\partial t)^2$ represents the influence of the compressibility of the air in the stability criteria.

14.6: Show in detail the steps between (14.115) and (14.118).

14.7: Divide equation (12.41) by M. Show that for the pseudoadiabatic process the factor dM'/M can be written as $m'dr^{21}/r^{21}$. Assume that the condensed water leaves the system immediately.

14.8: Derive an expression for γ_v as required in equation (14.86b). Assume the reversible moist adiabatic change of state and quasi-static conditions.

14.9: At the geopotential height h above the ground the temperature of the dry atmosphere assumes the value $T = T_0 \exp(-ah)$.

(a) Calculate the pressure $p(h)$ if the constant $a \neq 0$. Now assume that the constant a approaches zero to obtain the barometric height formula for an isothermal atmosphere.
(b) On the average, how many air molecules per unit volume n are found in the air column extending from the ground to the geopotential height h if $a = 0$.

14.10: Assume a linear temperature distribution in the height interval $(h_2 - h_1)$ as given by $T = T_0 - \gamma_g(h_2 - h_1)$ with $T(h = 0) = T_0$. If $(h_2 - h_1)$ is not too large, show that the arithmetic and the barometric mean temperatures are nearly identical. Hint: Discontinue the series expression of $\ln(1 + x)$ after the second term.

14.11: The temperature of the troposphere is assumed to change linearly with height from the ground $T(h = 0)$ to the tropopause height h^*. The stratosphere is characterized by the uniform temperature $T(h^*)$. For the dry atmosphere find a relation between the pressure tendency $\partial p_0/\partial t$ at the earth's surface and the height change of the tropopause $\partial h^*/\partial t$. Assume that the surface temperature T_0 does not change with time and that the pressure p_H at a sufficiently great height H remains constant.

Answers to problems

Chapter 1

1.1: (a) $l = 3$, (b) $l = 3/2$.

1.2: $l = 0$.

1.3: $v(p, T)$ is intensive $l = 0$ since p and T are intensive functions.

1.5: No special assumptions were made in the derivation of this equation. Thus, it is valid for open as well as closed systems.

1.7: $v_0 = R_0 T/p$, $v_1 = R_1 T/p$.

1.8: (b) Equations $u = u(x, y)$, $v = v(x, y)$ define the transformation between the points (u, v) of the (u, v)-plane and the points (x, y) of the (x, y)-plane. The inverse transformation is given by $x = x(u, v)$ and $y = y(u, v)$.

Chapter 2

2.1: The continuity equation $D\rho/Dt = 0$.

Chapter 3

3.4: The variation of e with respect to m^k, $k = 0, 1, 2, 3$ is overspecified.

3.5:

$$D = \left(\frac{\partial e}{\partial T}\right)_v \left(\frac{\partial p}{\partial v}\right)_T - \left(\frac{\partial p}{\partial T}\right)_v \left(\frac{\partial e}{\partial v}\right)_T$$

$$\left(\frac{\partial T}{\partial p}\right)_e = -\frac{1}{D}\left(\frac{\partial e}{\partial v}\right)_T, \qquad \left(\frac{\partial v}{\partial p}\right)_e = \frac{1}{D}\left(\frac{\partial e}{\partial T}\right)_v$$

$$\left(\frac{\partial T}{\partial e}\right)_p = \frac{1}{D}\left(\frac{\partial p}{\partial v}\right)_T, \qquad \left(\frac{\partial v}{\partial e}\right)_p = -\frac{1}{D}\left(\frac{\partial p}{\partial T}\right)_v$$

3.7:

$$\left(\frac{\partial e}{\partial v}\right)_T = 0$$

3.8: (a) exact, $f(x, y) = xy$; (b) not exact, $IF = 1/y^2$; (c) exact, $f(x, y) = (x^2 - y^2)/2$; (d) not exact, $IF = 1/x^2$.

3.9: $p = AT \exp[-a/(RvT)]/(v - b)$ with $A = (p_0/T_0)(v_0 - b)\exp[a/(Rv_0T_0)]$.

3.10:

(a) $\quad T(t) = \left(\dfrac{aT_0}{c_v} \displaystyle\int_0^t \exp(at'/c_v)v(t')^{R/c_v}dt' + T_i v_i^{R/c_v}\right)\exp(-at/c_v)v(t)^{-R/c_v}$

(b) $\quad \dfrac{T(t)}{T_i} = \left(\dfrac{v_i}{v(t)}\right)^{R/c_v}$ independent of T_0.

3.11:

$$e(T, v) = e(T_0, v_0) + c_v(T - T_0) + RT\{\exp[a/(vRT)] - \exp[a/(v_0RT)]\}$$

Chapter 4

4.1: (a)   In both cases 5.80 Joule K^{-1}. (b)   5.80 Joules K^{-1} for the gas and no entropy change for the universe.

4.2: -15.97 Joule K^{-1}.

4.3: $2Mc_p \ln[(T_1 + T_2)/2(T_1T_2)^{1/2}]$

4.6:

$$e = e_0 + \int_{T_0}^{T} c_v dT - \frac{a}{\breve{v}} + \frac{a}{\breve{v}_0}$$

$$s = \int_{T_0}^{T} c_v \ln T/T_0 + R^* \ln\left(\frac{\breve{v} - b}{\breve{v}_0 - b}\right)$$

Here $(\breve{v}_0, T_0)$ refer to reference values.

4.7: Answer: $q = R^*T \ln[(v_f - b)/(v_i - b)]$. v_f and v_i refer to the final and initial volume.

4.8: For the ideal gas: $(\partial e/\partial v)_T = 0$; for the van der Waals gas: $de = c_v dT + (a/\breve{v}^2)d\breve{v}$, $(\partial e/\partial \breve{v})_T = a/\breve{v}^2$.

4.10: (a)   2264 Joule g^{-1},   (b)   414 Joule g^{-1},   (c) -69.6 Joule g^{-1}.

Chapter 5

5.1: $Q = 4/3bT^4(V_f - V_i)$ where f and i stand for final and initial.

5.2: $C_v = 4bVT^3$.

5.3: $C_p = \infty$.

5.5: Due to (5.25) isothermal processes are also isobaric processes.

Chapter 6

6.6:

$$p = RT \left(\frac{\partial \ln z}{\partial v}\right)_T, \qquad\qquad s = R \ln Z + RT \left(\frac{\partial \ln Z}{\partial T}\right)_v$$

$$e = RT^2 \left(\frac{\partial \ln Z}{\partial T}\right)_v, \qquad\qquad h = RT^2 \left(\frac{\partial \ln Z}{\partial T}\right)_v + RTv \left(\frac{\partial \ln z}{\partial v}\right)_T$$

$$\mu = -RT \ln Z + RTv \left(\frac{\partial \ln Z}{\partial v}\right)_T$$

Chapter 7

7.4:

$$w_{\text{dif},2} = \frac{gL^{22}}{\rho^2} \left[\exp\left(\frac{-\rho^2}{L^{22}t}\right) - 1\right]$$

7.5:

$$T = \left[T_0 + \frac{g}{AR_0} \left(\frac{R_1}{R_0} - 1\right)\left(q_0 - \frac{K_1}{K_2}\right)\right] \exp\left(\frac{K_2 A}{K_1}z\right)$$
$$- \frac{g}{AR_0} \left(\frac{R_1}{R_0} - 1\right)\left(q_0 - Az - \frac{K_1}{K_2}\right)$$

7.6:

$$q = q_0 \exp\left[\frac{(R_1 - R_0)g}{R_0^2 T}z\right]$$

7.7:

$$m^1(z, t) = \sum_{n=1}^{\infty} D_n \sin\left(\frac{n\pi}{L}z\right) \exp\left[-\frac{D^{11}}{\rho}\left(\frac{n\pi}{L}\right)^2 t\right]$$
$$D_n = \frac{2}{L} \int_0^L f(z) \sin\left(\frac{n\pi}{L}z\right) dz, \qquad n = 1, 2, \ldots$$

Chapter 8

8.1: -1.528×10^6 J kg^{-1} mole^{-1}.

8.2:

$$s = s(T_0, p_0) + c_p \ln\left(\frac{T}{T_0}\right) - R \ln\left(\frac{p}{p_0}\right) - RA_1(p - p_0) - \frac{R}{2}A_2(p^2 - p_0^2)$$

$$v = \frac{RT}{p}(1 + A_1 p + A_2 p^2)$$

$$h = h(T_0, p_0) + c_p(T - T_0)$$

8.3:

$$p = \frac{RT}{v}\exp\left(\frac{a}{RTv}\right)$$

$$s = s(T_0, v_0) + \int_{T_0}^{T} \frac{\beta(T')}{T'}dT' - R[Ei(a/(RTv)) - Ei(a/(RT_0 v_0))]$$

$$+ R\left[\exp\left(\frac{a}{RTv}\right) - \exp\left(\frac{a}{RT_0 v_0}\right)\right]$$

$$e = e(T_0, v_0) + \int_{T_0}^{T} \beta(T')dT' + R\left[T\exp\left(\frac{a}{RTv}\right) - T_0\exp\left(\frac{a}{RT_0 v_0}\right)\right]$$

$$h = h(T_0, v_0) + \int_{T_0}^{T} \beta(T')dT' + 2R\left[T\exp\left(\frac{a}{RTv}\right) - T_0\exp\left(\frac{a}{RT_0 v_0}\right)\right]$$

$$\mu = \mu(T_0, v_0) - (T - T_0)s(T_0, v_0) + \int_{T_0}^{T} \beta(T')dT' - T\int_{T_0}^{T} \frac{\beta(T')}{T'}dT'$$

$$+ RT[Ei(a/(RTv)) - Ei(a/(RT_0 v_0))] + R(T - T_0)\exp\left(\frac{a}{RT_0 v_0}\right)$$

$$+ R\left[T\exp\left(\frac{a}{RTv}\right) - T_0\exp\left(\frac{a}{RT_0 v_0}\right)\right]$$

$$c_v = \beta(T) + \exp\left(\frac{a}{RTv}\right)\left(R - \frac{a}{Tv}\right)$$

8.4:

(a) $\check{S}_i = R^* \sum_k (\eta_k - \ln p)N^k$ with $\eta_k = \dfrac{\mathring{s}_k(p_0, T_0)}{R^*} + \dfrac{1}{R^*}\displaystyle\int_{T_0}^{T} \dfrac{\check{c}_{p,k}}{T'}dT'$

(b) $\check{S}_f = R^* \sum_k N^k(\eta_k - \ln p^k)$

(c) $\Delta S = -R^* \sum_k N^k \ln n^k$

(d) $\Delta\check{S} = 2R^* \ln 2 > 0$

The subscripts i, f refer to initial and final, respectively, the overhead symbol ˘ indicates that the quantities refer to mole and not to unit mass. $\dot{s}_k$ is a combination of constants.

8.5:

$$s = s(0) + c_v \ln\left(\frac{T}{T_0}\right) + R \ln\left(\frac{v - \alpha}{v_0 - \alpha}\right)$$

$$p = \frac{RT}{v - \alpha} - \frac{\beta}{v^2}$$

$$e = e(0) - \beta\left(\frac{1}{v} - \frac{1}{v_0}\right) + c_v(T - T_0)$$

$$c_p = c_v + \frac{R^2 T v^3}{RT v^3 - 2\beta(v - \alpha)^2}$$

Chapter 10

10.1: (a) 155×10^3 J,   (b) 450 J K^{-1}.

10.2: (a) $\check{v}_{cr} = 3b$,   (b) $T_{cr} = \frac{8a}{27bR^*}$,   (c) $p_{cr} = \frac{a}{27b^2}$

10.4: 2820×10^3 J kg^{-1}.

10.6: (a) $\approx 70°$C,   (b) ≈ 23 km.

10.7:

$$T_f = T_i - \frac{l_{21}(T_i)}{c_{p,0}m^0 + c_{p,1}m^{H_2O}}, \qquad m^{H_2O} = m^1 + m^2$$

10.8: $\Delta T_{melt} = -0.73$ K.

Chapter 12

12.1:

$$\left(\frac{\partial T}{\partial p}\right)_{\theta_h} = \frac{k_0 T}{p} \frac{1 + \frac{l_{21} r^{21}}{R_0 T}}{1 + \frac{l_{21}^2 r^{21}}{c_{p,0} R_1 T^2}}$$

Chapter 13

13.2: $T_{e,m} \approx 287$ K.

Chapter 14

14.2: 40 gpm.

14.3: (b)   8000 gpm.

14.4: For dry adiabatic stratification $n = (1-k_0)/k_0$ so that $p = p_0(1-h/H_a)^{(1/k_0)}$ with $k_0 = R_0/c_{p,0}$, $H_a = H_{p,0}/k_0$, and $\rho = \rho_0(p/p_0)^{(1-k_0)}$.

14.8:

$$\gamma_v = \frac{T_v}{T}\left(1 + q^{21}\frac{R_1 - R_0}{R_0}T\frac{d\ln p^{21}}{dT}\right)\gamma - \frac{9.8}{R_0}\frac{R_1 - R_0}{R_0}q^{21}$$

where q^{21} is the saturation specific humidity. The same answer applies to the irreversible change of state.

14.9: (a)

$$p = p_0\exp\left(-\frac{9.8}{R_0 T_0}\frac{\exp(ah)-1}{a}\right),$$

$$a \longrightarrow 0: \qquad p = p_0\exp\left(-\frac{9.8h}{R_0 T_0}\right)$$

(b)

$$n = n_0\frac{H_p}{h}\left[1 - \exp(-h/H_p)\right] \quad \text{with} \quad H_p = \frac{R_0 T_0}{9.8}$$

n_0 is the number of molecules at $h = 0$.

14.11:

$$\frac{\partial p_0}{\partial t} = \frac{p_0 9.8}{R_0}\frac{\gamma_g(H - h^*)}{(T_0 - \gamma_g h^*)^2}\frac{\partial h^*}{\partial t}$$

Frequently used symbols

a_{21}:	chemical affinity of vaporization	$(\mathrm{m^2\ s^{-2}})$
a_{31}:	chemical affinity of sublimation	$(\mathrm{m^2\ s^{-2}})$
a_{32}:	chemical affinity of melting	$(\mathrm{m^2\ s^{-2}})$
c_p:	specific heat at constant pressure	$(\mathrm{m^2\ s^{-2}\ K^{-1}})$
c_v:	specific heat at constant volume	$(\mathrm{m^2\ s^{-2}\ K^{-1}})$
D/Dt:	budget operator	$(\mathrm{s^{-1}})$
$đA$:	amount of work	$(\mathrm{kg\ m^2\ s^{-2}})$
$đa$:	specific amount of work	$(\mathrm{m^2\ s^{-2}})$
$đQ$:	amount of heat	$(\mathrm{kg\ m^2\ s^{-2}})$
$đq$:	specific amount of heat	$(\mathrm{m^2\ s^{-2}})$
$d\mathbf{S}$:	surface element	$(\mathrm{m^2})$
$d\tau$:	volume element	$(\mathrm{m^3})$
E:	internal energy	$(\mathrm{kg\ m^2\ s^{-2}})$
e:	specific internal energy	$(\mathrm{m^2\ s^{-2}})$
$\mathbb{E}$:	unit dyadic	
F:	free energy	$(\mathrm{kg\ m^2\ s^{-2}})$
f:	specific free energy	$(\mathrm{m^2\ s^{-2}})$
$\mathbf{F}_R$:	radiative flux	$(\mathrm{kg\ s^{-3}})$
G:	chemical potential, free enthalpy, Gibbs function	$(\mathrm{kg\ m^2\ s^{-2}})$
$\mathbf{g}$:	gravity vector	$(\mathrm{m\ s^{-2}})$
g:	acceleration of gravity	$(\mathrm{m\ s^{-2}})$
H:	enthalpy	$(\mathrm{kg\ m^2\ s^{-2}})$
H_p:	pressure scale height	$(\mathrm{m^2\ s^{-2}})$
h:	specific enthalpy	$(\mathrm{m^2\ s^{-2}})$
h:	geopotential height	$(\mathrm{m^2\ s^{-2}})$
I^k:	phase transition rate of substance k	$(\mathrm{kg\ m^{-3}\ s^{-1}})$
$\mathbb{J}$:	viscous stress tensor	$(\mathrm{kg\ m^{-1}\ s^{-2}})$

J^h:	enthalpy flux	(kg s^{-3})
J^h_s:	sensible enthalpy flux	(kg s^{-3})
J^k:	diffusion flux of substance k	$(\text{kg m}^{-2} \text{ s}^{-1})$
k:	specific kinetic energy	$(\text{m}^2 \text{ s}^{-2})$
k_b:	specific kinetic energy of barycentric motion	$(\text{m}^2 \text{ s}^{-2})$
k_d:	specific kinetic energy of diffusion motion	$(\text{m}^2 \text{ s}^{-2})$
l_{21}:	latent heat of vaporization	$(\text{m}^2 \text{ s}^{-2})$
l_{31}:	latent heat of sublimation	$(\text{m}^2 \text{ s}^{-2})$
l_{32}:	latent heat of melting	$(\text{m}^2 \text{ s}^{-2})$
L^{ij}:	phenomenological coefficient	
l^{11}:	coefficient of volume viscosity	$(\text{kg m}^{-1} \text{ s}^{-1})$
l^{ij}:	phenomenological coefficient	
l_m:	permeability coefficient	(s)
l_{mq}:	coefficient of thermal osmosis	$(\text{kg m}^{-1} \text{ s}^{-1} \text{ K}^{-1})$
l_{qm}:	coefficient of osmotic flow	$(\text{m}^2 \text{ s}^{-1})$
l_q:	heat conduction coefficient	$(\text{kg m s}^{-3} \text{ K}^{-1})$
M:	mass	(kg)
M^k:	partial mass of substance k	(kg)
m^k:	concentration of substance k	
$\breve{m}_k$:	molecular weight of substance k	(kg mole^{-1})
$\mathbf{N}$:	solenoidal vector	
N^k:	number of moles of substance k	
n^k:	molar fraction of substance k	
p:	pressure	$(\text{kg m}^{-1} \text{ s}^{-2})$
p^k:	partial pressure of substance k	$(\text{kg m}^{-1} \text{ s}^{-2})$
p^{k1}:	saturation vapor pressure	$(\text{kg m}^{-1} \text{ s}^{-2})$
p^{32}:	saturation melting pressure	$(\text{kg m}^{-1} \text{ s}^{-2})$
Q_s:	entropy production	$(\text{kg m}^{-1} \text{ s}^{-3} \text{ K}^{-1})$
Q_ψ:	source term of quantity ψ	
q:	specific humidity	
R:	individual gas constant	$(\text{m}^2 \text{ s}^{-2} \text{ K}^{-1})$
R^*:	universal gas constant	$(\text{kg m}^2 \text{ s}^{-2} \text{ K}^{-1} \text{ mole}^{-1})$
R_0:	gas constant of dry air	$(\text{m}^2 \text{ s}^{-2} \text{ K}^{-1})$
R_1:	gas constant of water vapor	$(\text{m}^2 \text{ s}^{-2} \text{ K}^{-1})$
R_m:	gas constant of moist air	$(\text{m}^2 \text{ s}^{-2} \text{ K}^{-1})$
r:	mixing ratio	
r^{k1}:	saturation mixing ratio	
$\mathbf{r}$:	position vector	
S:	entropy	$(\text{kg m}^2 \text{ s}^{-2} \text{ K}^{-1})$
s:	specific entropy	$(\text{m}^2 \text{ s}^{-2} \text{ K}^{-1})$

T:	temperature	(K)
T_d:	dew point temperature	(K)
T_e:	equivalent temperature	(K)
T_v:	virtual temperature	(K)
t:	time	(s)
U:	relative humidity	
V:	volume	(m^3)
V^k:	volume of substance k	(m^3)
v:	specific volume	$(m^3\,kg^{-1})$
v_k:	specific volume of substance k	$(m^3\,kg^{-1})$
v^k:	partial specific volume of substance k	$(m^3\,kg^{-1})$
$\mathbf{v}$:	barycentric velocity	$(m\,s^{-1})$
$\mathbf{v}_k$:	velocity of substance k	$(m\,s^{-1})$
$\mathbf{v}_{k,d}$:	diffusion velocity of substance k	$(m\,s^{-1})$
v^*:	isobaric expansion coefficient	(K^{-1})
$đW$:	amount of work	$(kg\,m^2\,s^{-2})$
$đw$:	specific amount of work	$(m^2\,s^{-2})$
β:	isochoric pressure coefficient	(K^{-1})
Γ_p^ρ:	coefficient of barotropy	$(s^2\,m^{-2})$
γ_p^ρ:	piezotropy coefficient	$(s^2\,m^{-2})$
γ:	general symbol for lapse rate	$(K\,gpm^{-1})$
ϵ:	specific total energy	$(m^2\,s^{-2})$
κ:	isothermal compressibility	$(kg^{-1}\,m\,s^2)$
κ_s:	adiabatic compressibility	$(kg^{-1}\,m\,s^2)$
μ:	specific Gibbs function, specific free enthalpy	$(m^2\,s^{-2})$
μ_k:	chemical potential	$(m^2\,s^{-2})$
ν:	coefficient of shear viscosity	$(kg\,m^{-1}\,s^{-1})$
Π:	Exner function	$(m^2\,s^{-2}\,K^{-1})$
ρ:	mass density	$(kg\,m^{-3})$
ρ^k:	partial density of substance k	$(kg\,m^{-3})$
σ_0:	static stability	$(kg^{-2}\,m^4\,s^2)$
θ:	potential temperature	(K)
θ_e:	potential equivalent temperature	(K)
θ_l:	liquid water potential temperature	(K)
θ_{ps}:	pseudopotential temperature	(K)
ϕ:	geopotential	$(m^2\,s^{-2})$
ϕ_a:	attractional potential	$(m^2\,s^{-2})$
ϕ_c:	centrifugal potential	$(m^2\,s^{-2})$
$\overset{\circ}{\psi}_k$:	specific value of Ψ of substance k in the pure phase	
$\breve{\psi}$:	molar value of Ψ	

Constants

c_L:	Laplace speed of sound at 273 K	(331 m s^{-1})
γ:	gravitational constant	$(6.672 \times 10^{-11} \text{ Nm}^2 \text{ kg}^{-2})$
$c_{p,0}$:	specific heat at constant pressure, dry air	$(1005 \text{ J kg}^{-1} \text{ K}^{-1})$
$c_{p,1}$:	specific heat at constant pressure, water vapor	$(1847 \text{ J kg}^{-1} \text{ K}^{-1})$
$c_{v,0}$:	specific heat at constant volume, dry air	$(718 \text{ J kg}^{-1} \text{ K}^{-1})$
$c_{v,1}$:	specific heat at constant volume, water vapor	$(1386 \text{ J kg}^{-1} \text{ K}^{-1})$
c_2:	specific heat of liquid water	$(4190 \text{ J kg}^{-1} \text{ K}^{-1})$
c_3:	specific heat of ice	$(2090 \text{ J kg}^{-1} \text{ K}^{-1})$
l_{21}:	latent heat of vaporization	$(2.5 \times 10^6 \text{ J kg}^{-1})$
l_{31}:	latent heat of sublimation	$(2.834 \times 10^6 \text{ J kg}^{-1})$
l_{32}:	latent heat of melting	$(0.334 \times 10^6 \text{ J kg}^{-1})$
M_E:	mass of the earth	$(5.973 \times 10^{24} \text{ kg})$
R^*:	universal gas constant	$(8.31432 \text{ J mole}^{-1} \text{ K}^{-1})$
R_0:	gas constant of dry air	$(287.05 \text{ J kg}^{-1} \text{ K}^{-1})$
R_1:	gas constant of water vapor	$(461.51 \text{ J kg}^{-1} \text{ K}^{-1})$
σ:	Stefan–Boltzmann constant	$(5.67 \times 10^{-8} \text{ J m}^2 \text{ s}^{-1} \text{ K}^{-4})$
T_{trip}:	temperature of the triple point	(273.16 K)

References and bibliography

Betts, A. K., 1973: Non-precipitating cumulus convection and its parameterization. *Q. J. R. Meteorol. Soc.*, **99**, 178–196.

Bjerknes, V., J. Bjerknes, H. Solberg, and T. Bergeron, 1933: *Physikalische Hydrodynamik*. Springer Verlag, New York.

Bjerkness, J., 1938: Saturated ascent of air through a dry-adiabatically descending environment. *Q. J. R. Meteorol. Soc.*, **65**, 325–330.

Bohren, C. F., and B. A. Albrecht, 1998: *Atmospheric Thermodynamics*. Oxford University Press, New York; Oxford.

Callies, U., and F. Herbert, 1984: *On the Treatment of Radiation in the Entropy Budget of the Earth–Atmosphere System. New Perspectives in Climate Modelling*. Elsevier Science Publ. B. V., Amsterdam, The Netherlands, pp. 311–329.

Carslaw, H. S., and G. C. Jaeger, 1959: *Conduction of Heat in Solids*. Clarendon Press, Oxford, UK.

Casimir, H. B. G., 1945: On Onsager's principle of microscopic reversibility. *Rev. Mod. Phys.*, **17**, 343–350.

Chapman, S., and T. G. Cowling, 1939: *The Mathematical Theory of Non-uniform Gases*. Cambridge University Press, Cambridge, UK.

Curry, J. A., and P. J. Webster, 1999: *Thermodynamics of Atmospheres and Oceans*. Academic Press, Orlando, FL, USA.

De Groot, S. R., 1960: *Thermodynamik irreversibler Prozesse*. Hochschultaschenbücher Verlag, Bibliographisches Institut, Mannheim, Germany.

Dinkelacker, O., 1939: Die Feuchtadiabate thermodynamisch irreversibler Prozesse. *Meteor. Zeitschrift*, **56**, 289–297.

Eckart, C., 1940: The thermodynamics of irreversible processes I. The simple fluid. *Phys. Review*, **58**, 267–269.

Haltiner, G. J., and F. L. Martin, 1957: *Dynamical and Physical Meteorology*. McGraw-Hill Book Company, Inc., New York; Toronto; London.

Hauf, T., and H. Höller, 1987: Entropy and potential temperature. *J. Atmos. Sci.*, **43**, 2887–2901.

Herbert, F., 1973: Irreversible processes in the atmosphere, part 2. *Beitr. Phys. Atmos.*, **46**, 262–288.

Herbert, F., and J. Pelkowski, 1990: Radiation and entropy. *Beitr. Phys. Atmos.*, **63**, 134–140.

Iribarne, I. V., and W. L. Godson, 1981: *Atmospheric Thermodynamics*. D. Reidel, Dordrecht, The Netherlands.

Kluge, G., and G. Neugebauer, 1976: *Grundlagen der Thermodynamik*. VEB Deutscher Verlag der Wissenschaften, Berlin, Germany.

Lee, J. F., F. W. Sears, and D. Turcotte, 1963: *Statistical Thermodynamics*. Addison-Wesley Publishing Company, Inc. Reading, MA; Palo Alto; London.

Margules, M., 1905: Über die Energie der Stürme. J. B. K. K. Zentralanstalt für Met. und Erdmagn., Wien, 40. Bd., 1903–1921.

Onsager, L., 1931a: Reciprocal relations in irreversible processes. I. *Phys. Rev.*, **37**, 405–426.

Onsager, L., 1931b: Reciprocal relations in irreversible processes. II. *Phys. Rev.*, **38**, 2265–2279.

Päsler, M., 1975: *Phänomenologische Thermodynamik*. Walter de Gruyter, Berlin; New York.

Reichel, L. E., 1980: *A Modern Course in Statistical Physics*. Edward Arnold Publishers Ltd, London, UK.

Showalter, A. K., 1953: A stability index for thunderstorm forecasting. *Bull. Am. Meteorol. Soc.*, 34, 250–252.

Sommerfeld, A., 1965: *Vorlesungen über Theoretische Physik, Bd. V, Thermodynamik und Statistik*. Akademische Verlagsgesellschaft, Geest and Portig, Leipzig, Germany.

United States Air Force, 1961: *Use of the Skew T-Log p Diagram in Analysis and Forecasting*.

Van Mieghem, J., 1973: *Atmospheric Energetics*. Oxford Monographs on Meteorology, Clarendon Press, Oxford, UK.

Zdunkowski, W., and T. Kandlbinder, 1997: An analytic solution to nocturnal cooling. *Beitr. Phys. Atmos.*, **69**, 337–348.

Zemansky, M., 1968: *Heat and Thermodynamics*. McGraw Hill Book Company, Inc. New York; Toronto; London.

Index